城市设计研究

城中村的新生

REBIRTH OF THE URBAN VILLAGE

杨镇源 肖　靖 (深圳大学)
李凌月 许　凯 孙彤宇 (同济大学)
唐　斌 (东南大学)
[奥] 莫拉登·亚德里奇
(Mladen Jadric)
(维也纳技术大学)　著

中国城市出版社

关于本书

深圳是我国人口密度最高的城市，也是城镇率达100%的城市。由于新增土地供应不足，城市更新逐渐成为深圳城市建设发展的主导模式，城市从增量发展时代转变为存量发展时代。深圳还有大量的城中村将被拆除，或者被改造。城市设计作为管控城市发展的重要手段，也必须去适应这个变化，对城中村保留改造的可行性进行研究，针对改造提出设计策略和指导方针。

本书收录了“中奥四校联合城市设计营”针对这个课题所做的研究和设计。特别感谢参与此次工作营的所有老师、同学，还有参与指导评图的各位嘉宾，同时感谢同济大学、东南大学还有维也纳技术大学的支持与帮助，使得本书能以城市设计研究系列丛书的方式出版。

ABOUT THIS BOOK

Shenzhen is the city with the highest population density in China, and also is the city with 100% completion of urbanization. Due to the lack of land supply, urban renewal has gradually become the leading mode of urban and development in Shenzhen, and the city has changed from the era of incremental development to the era of stock development. A large number of villages in Shenzhen will be demolished or rebuilt. As an important mean to control urban development, urban design must also adapt to this change, study the feasibility of urban village reservation and transformation, and put forward design strategies and guidelines for the transformation.

This book contains the research and design of “The Austrian-Chinese Joint Design Camp” for this subject. Special thanks to all the teachers and students who participated in the camp, as well as all the guests who participated in the guidance and critic of the workshop. At the same time, thanks to the support and help of Tongji University, Southeast University and TU Vienna, this book can be published as a series of urban design research books.

关于“2018中奥四校联合城市设计营”

“中奥四校联合城市设计营”由维也纳技术大学、同济大学、东南大学和深圳大学联合创办于2002年。每年夏天由四所学校轮流作为东道主来选题和组织，至今已成功举办十余次。设计工作坊关注中西方近十年来共同面临的城市问题：如产业区改造、旧城更新、历史街区活化、新城规划、新农村建设等；通过十多年的积累与发展，从仅有建筑学专业参与开始扩展，通过不同专业背景，跨区域的思维碰撞，研究创新的、带有一定实验性的城市设计策略。通过15天左右的短期交流与设计，汇聚四校学生与老师的创意与努力,形成工作坊的设计研究成果。

2018年度的中奥四校联合城市设计营落户深圳大学,以“城中村的新生”为课题,研究城中村的改造问题,探索除拆除重建以外的城市更新办法，课题选取了南山区的麻磡村作为城市设计的课题基地。

About “Austrian-China Joint Design Camp 2018”

“The Austrian-Chinese Joint Design Camp”was established by TU Vienna,Southeast University,Shenzhen University and Tongji University in 2002,and has been hosted by each university annually since then.The Camp takes the key urban issues that involve both Austria and China in recent years,such as industrial area conversion, urban regeneration, historic area revitalization as well as pedestrian and public space planning, as themes of design.In a term of 10 days, professors and students from four universities devote their creativity and effort towards diverse solutions to such issues.

In 2018,the Austrian-Chinese Joint Design Camp took place in Shenzhen University.Taking the “Rebirth of the urban villages” as the subject, the workshop studies the transformation of the village in the city, and selects the Makan village in Nanshan District as the project base for urban design.

序
PREFACE

麻磡村，孕育于山环水抱之间，发祥于中西文化初融之世，再兴盛于全球化的产业转移浪潮之际，近年来，又因深圳新的高等教育集群布局于此而变得格外引人注目。然而由于地处深圳市的重要水源地上游，且全域处于生态保护红线内，使得麻磡村的新机遇带着独特的老约束，并与深圳众多“城边村”一起，成为深圳市空间规划和设计中一个矛盾日益尖锐的课题。

“城边村”的故事不仅带着所有城中村的历史纠葛、主体多元、身份模糊和非正规性的问题，而且承载了深港双城的生存血脉，这使得他们在经历了开放早期的无序扩张之后，已基本失去了空间再度膨胀的机会。因此，基于“利益平衡”基础上的“生态保护”和“资源再生”成为城中村发展尝试的关键。麻磡村目前更多的是面向高校集群的“创新服务”。本次四校联合设计工作坊从建筑学的视角拓展到社会网络的关系，全面审视麻磡村可持续发展的多种可能性。

利益平衡: 城市更新是动态的发展，需要不同的利益主体共同协作完成。在本次工作坊中，参与小组尝试以“自下而上”的设计思想指导学生进行设计构思。研究在城市更新过程中，各个不同利益主体的诉求、参与方式及对片区更新的影响。从设计策略、开发建设、投资运营等多个方面，建立相互联系的整体城市设计方法理论研究。

生态保护: 尽管生态保护红线的存在限制了麻磡村“野蛮式”的扩张，但却未能阻止多年来村内厂房排放所造成的环境污染。参与工作坊的团队，由建筑、城乡规划、风景园林多个专业背景的成员组成，提出了生态系统重构的工作计划，以修复水循环系统的方式对片区生态环境提出实质性的策略。

资源再生: 要发展麻磡，同时要避免破坏生态，最好的办法就是依托现有资源进行更新再用。将原有破败的各类资源，包括废弃厂房、空置民宅、荒地进行改造，重新利用，变废为宝。工作坊的设计团队对现有的“闲置”空间资源进行了重新定位，结合其空间特点提出针对性的活化策略。这些策略之中，包括大开脑洞的“城市农村”模式，对生态与产业结合的发展模型进行了探索。

创新服务: 麻磡村所处的南山区是我国创新经济的首善之区，2km范围内高校集群区的谋划落地更给麻磡村的发展带来无限遐想。近年来，深圳老旧社区中逐渐萌生的创新创意产业，正在一点点成功重塑城市环境。工作坊团队提出的基于“自组织理论”作用下的改造模式，为空间受限的麻磡村进行创新服务空间的转型升级改造提出了多种可能的解决方案。

工作坊以国际跨校联合的共创模式，在深圳的城边村操演了一次综合性空间思维训练，相信各方院校都受益匪浅，师生们均格外珍惜交流与碰撞所带来的知识与创新。这项由同济大学、东南大学、维也纳技术大学和深圳大学联合发起的中奥四校联合教学活动，已历经十年，非常难得，衷心期待国际教学科研合作的道路越走越好。

杨晓春

深圳大学建筑与城市规划学院　教授、副院长、规划系主任

2020 年 5 月 23 日于深圳大学

Makan village was born between mountains and rivers, originated in the early fusion of Chinese and Western cultures, and flourished in the wave of global industrial transfer. However, it is located in the upstream of the important water source of Shenzhen City, and the whole area is within the red line of ecological protection, which makes the new opportunities of Makan village with unique old constraints. Together with many "City-edge villages" in Shenzhen, it has become an increasingly contradictory topic in the spatial planning and design of Shenzhen city.

The story of "City-edge village" not only has the historical entanglement, subject diversity, identity ambiguity and informality of all the villages in the city, but also carries the surviving blood of Shenzhen and Hong Kong. It also makes them lose the opportunity of space expansion after the disordered expansion in the early stage of Opening-up. Therefore, "ecological protection" and "resource regeneration" based on the "balance of interests" have become the key words for the development of urban-side villages, and Makan village has more "innovative services" for university clusters. The joint design workshop of the four schools expanded from the perspective of architecture to the relationship of social network to comprehensively examine the various possibilities of sustainable development of Makan village.

Balance of interests: urban renewal is a dynamic development, involving the cooperation of different stakeholders. In this workshop, the participating group tried to guide the students to design ideas with bottom-up design ideas. In the process of urban renewal, the demands of different stakeholders, the ways of participation and the impact on regional renewal are studied.
Ecological protection: Although the existence of the red line of ecological protection limits the "barbaric" expansion of the village, it fails to prevent the environmental pollution caused by the emission of factories in the village for many years. The team participating in the workshop puts forward the work plan of eco-system reconstruction, and puts forward substantive strategies for the regional ecological environment in the way of repairing the water circulation system.

Resource regeneration: to develop and avoid ecological damage, the best way is to rely on existing resources for renewal and reuse. All kinds of resources, including abandoned factories, vacant houses and wasteland, will be transformed, reused and turned into treasures. The design team of the workshop repositioned the existing "idle" space resources, and put forward targeted activation strategies based on its space characteristics.

Innovation Service: the Nanshan District, where Makan village is located, is the leading area of innovation economy in China. The planning and implementation of the university cluster area within 2km brings infinite reverie to the development of Makan village. In recent years, the innovative and creative industries emerging from the old communities in Shenzhen are gradually reshaping the urban environment.

Teachers and students cherish the knowledge and innovation brought by exchange and interaction. This joint teaching activity of the four universities of China and Austria, jointly sponsored by Tongji University, Southeast University, Vienna University of Technology and Shenzhen University, has gone through ten years, which is very precious. I sincerely hope that the international teaching and research cooperation will be better and better.

Prof. YANG Xiaochun

Deputy dean, SAVP, head of urban planning department
May 23th, 2020, Shenzhen University

研究团队
RESEARCH TEAM

杨镇源 | 深圳大学建筑与城市规划学院讲师，国家一级注册建筑师，英国注册建筑师，英国皇家建筑师协会会员。曾获“德意志学术交流大学”（DAAD)的艺术家奖金（建筑类）赴德留学，其后在欧洲UNstudio、Foster+Partners等知名事务所工作多年。2011年回国后在深圳大学任教，并创办拾陌设计工作室，主要从事参数化设计和城市设计等相关研究领域的教学、研究与创作。

Mr.YANG Zhenyuan,Class1PRC,ARB,RIBA, received his B.Arch. from Shenzhen University.He was later awarded the DAAD artist scholarship that supported his study at Frankturt Academy of Fine Arts(Steadelschule)tutored by Dutch architect Ben Van Berkel to complete his master. After years of experience working in Europe for well-recognized design firms, including UN studio and Foster+Partners,he is now teaching at the college of architecture and urban planning,Shenzhen University. His main research scopes are related to parametric design and urban design.

肖靖 | 深圳大学建筑与城市规划学院助理教授，英国诺丁汉大学建筑与建成环境学院博士，曾获国际U21大学联盟奖学金赴加拿大麦吉尔大学做访问博士生。博士论文和近期研究领域包括建筑知识与图像媒介的近代转型及其相关建筑与城市发展史，已在诸多国际专业学术期刊发表专题文章。

Dr Jing Xiao works as an assistant professor in architecture at Shenzhen University, China. He received a PhD from the University of Nottingham in 2013 with his dissertation focusing on the transformation of architectural knowledge via pictorial media in pre-modern China. He publishes refereed papers in the field of pre-modern architectural history and modern urban morphology in many international academic journals including Open House International, Studies in the History of Gardens & Designed Landscapes, IDEA Journal and Habitat International.

李凌月 | 同济大学建筑与城市规划学院助理教授。先后毕业于同济大学、香港大学，获香港大学博士学位。主要研究方向包括城市治理、城市政策与理论、城市更新与城市设计。曾任职于上海市城市规划设计研究院、香港大学城市研究中心、上海市发展改革研究院。获香港大学杰出研究型毕业生，入选上海市浦江人才，上海市发展改革经济学论坛个人奖。先后主持参与国家自然科学基金，科技部、上海市等多项国家、省部级科研项目，以及多项咨询项目。

Lingyue LI is currently an assistant professor of College of Architecture and Urban Planning (CAUP), Tongji University. She received her B.E. degree in urban planning from Tongji University, M.Phil. and Ph.D. degrees from DUPAD, HKU. Dr. LI's research area lies in urban studies and urban planning, with a focus on theory and practice in urban China transition, urban regeneration and design, and urban innovation. Dr. LI is also awarded with “Outstanding Postgraduate (HKU)” and “Shanghai Pujiang Program”. She has been PI for research projects sponsored by National Natural Science Foundation and has participated in a number of state- and provincial-level projects funded by the Ministry of Science and Technology, Shanghai Development and Reform Commission, Shanghai Science and Technology Commission.

许凯 | 同济大学建筑与城市规划学城市设计团队副教授，同济—维也纳工大双学位项目协调人。同济大学建筑与城市规划学院硕士，维也纳工大建筑学院博士，师从克劳斯·泽姆斯罗特教授。主要研究领域是城市设计与产业空间规划。作为上海尤根建筑设计的创办者，从事建筑设计和城市设计方面的工作。

Dr. Xu Kai is an associate professor in Tongji University of Shanghai. He obtained his doctor degree in University of Technology,Vienna. Prof. Klaus Semsroth was his tutor professor. Dr. Xu's research areas include urban design and industrial space planning. Being the founder of Jugend Architecture, he also actively practice in architecture design and urban design.

孙彤宇 | 同济大学建筑与城市规划学院副院长、教授、博士生导师，中国建筑学会城市设计分会常务理事，建筑教育评估分会副理事长，中国建筑学会标准工作委员会委员，上海市绿色建筑协会副会长。同济大学建筑与城市规划学院博士，德国柏林工大和斯图加特大学访问学者。主要研究领域为城市设计及建筑设计理论与方法，主持多项国家和省部级科研项目。同时也是活跃的从业建筑师，是同济大学建筑设计研究院（集团）有限公司都市建筑设计院四所所长和主创建筑师，有多项设计获国家和地方建筑设计奖。

Prof. Dr. Sun Tongyu obtained his doctor degree in Tongji University in Shanghai. He was the visiting scholar in TU Berlin in 2003, and in Stuttgart University in 2008. He is now the vice dean of the College of the Architecture and Urban Planning in Tongji University, executive director of urban design branch, vice chairman of architecture education evaluation branch, member of code constitution branch of Architectural Society of China, he is also the executive director of the Green Building Society of Shanghai. His main research filed is urban design and architecture design. He leads various national and provincial-level research projects, and also actively practices in architecture design and urban design. He is the director of the Design Division No. 4 in Tongji Design Institute.

唐斌 | 东南大学建筑学院副教授，曾任职于南京军区建筑设计院，东南大学建筑设计与理论研究中心，国家一级注册建筑师。主要从事建筑设计、城市设计相关理论研究与设计创作。主要研究方向：高密度传统街区更新方法研究；基于地域性与现代性的当代乡村营造技术与方法；渐进式城市设计理论与方法。先后主持和参与20余项工程设计项目，多次获得省市级优秀工程设计奖，并在重要的设计竞赛与投标中获胜。教学方面，主要负责东南大学建筑学院本科三年级及硕士研究生的建筑设计课教学，多次作为指导教师获得高等级学生竞赛奖项。

Tang Bin, associate professor of School of architecture of Southeast University, once worked in Nanjing Military Region Architectural Design Institute, architectural design and theory research center of Southeast University National first-class registered architect. Mainly engaged in architectural design, urban design related theoretical research and architectural design creation. The main research directions include: Research on high-density traditional street renewal methods; Contemporary rural construction technology and methods based on regionality and modernity; Progressive urban design theory and methods. He has presided over and participated in more than 20 architectural design projects, won many provincial and municipal Excellent Architecture Designing Awards, and won important design competitions and bids. In terms of teaching, he is mainly responsible for the teaching of architectural design course for the third-year undergraduate and master's degree students in the school of architecture of Southeast University, and has won high-level students' competition award as the instructor for many times.

莫拉登·亚德里奇（奥）（Mladen Jadric） | 任教于维也纳技术大学，维也纳技术大学中奥交流项目负责人，奥地利艺术家协会成员，维也纳建筑与工程师协会副会长。自1997年起，他开始在维也纳工科大学任教，并担任欧洲、北美、亚洲、澳洲和南美洲等多所学校的客座教授，其作品在世界多地展出，包括英国皇家艺术学院、美国麻省理工学院、库珀联盟、罗杰威廉姆斯大学、芬兰阿尔托大学、威尼斯双年展、东京世界建筑展、柏林20世纪博物馆、首尔建筑双年展等。他曾经获得多个国家级的建筑设计奖，包括维也纳住宅奖，韩国釜山市长奖等。

Mladen Jadric, teaching and practicing architecture in vienna, Austria. He is head of exchange programs and workshops with Asia and double-degree program with Tongji University, Shanghai,China. Since 1997 he has been teaching at the Faculty of Architecture and Planning (TU Wien) and has gained extensive experience as a visiting professor and guest lecturer in Europe, USA, Asia, Australia, and South America. He is a member of Vienna Künstlerhaus. Since 2018 he is a deputy section chairman of architects of the Chamber of Architects and Engineers of Vienna, Lower Austria and Burgenland. His work has been exhibited at the Royal Academy of Arts in London, UK; M.I.T., Cooper Union, and Roger Williams University, USA; Alvar Aalto University, Helsinki, Finland; the Architectural Biennale in Venice, Italy; the World Architectural Triennale in Tokyo, Japan; Museum of the 20th Century, Berlin, Germany; Seoul Biennale of Architecture and Urban Planning, Korea; NIT-Nagoya Institute of Technology in Japan, and many more. He received the State Award for Experimental Architecture, Karl Scheffel Preis and Schorsch Preis for housing by City of Vienna. Grand Prize by Mayor of Busan Metropolitan City, Republic of Korea a.o. Member of “Künstlerhaus”, oldest association of artists and architects in Austria.

研究参与人员

RESEARCH PARTICIPANTS

维也纳技术大学 TU Wien:

Jääskeläinen Veera Sofia, Kacar Emre, Kitzberger Fabian, König Severin , Mate Andras, Milojevic Aleksa, Mirzaiyan-Tafty Marco,Pakfiliz Pinar.

同济大学 Tongji University:

王浩宇 Wang Haoyu, 孙泽龙 Sun Zelong, 张黎晴 Zhang Liqing, 朱薛景 Zhu Xuejing, 乔映荷 Qiao Yinghe, 林妮 Lin Ni, 白智成 Bai Zhicheng.

东南大学 SouthEast University:

郭浩然 Guo Haoran, 叶津岑 Ye Jincen, 王曹寿轩 Wang Caoshouxuan, 续爽 Xu Shuang, 徐海琳 Xu Hailin, 王家琳 Wang Jialin.

深圳大学 Shenzhen University:

史心悦 Shi Xinyue, 张红连 Zhang Honglian, 何巧 He Qiao, 夏浩桐 Xia Haotong, 刘智仪 Liu Zhiyi, 李庄庭 Li Zhuangting, 胡平 Hu Ping, 刘真鑫Liu Zhenxin, 李嘉芙Li Jiafu

目录

CATALOGUE

城中村与城市的关系暧昧，一方面它自成体系，由村集体股份公司管理；另一方面它为全市提供了一半的居住空间，有上千万城市人口常住在村里。伴随着经济发展，城中村不再与城市对立，反而形成一种补充。村和城以共生的形式存在，彼此不可分离。由于这种变化，本该“拆除”的城中村，通过“改造”的方式来适应周边城市的发展。

The relationship between the city and urban village is vague. On the one hand, urban village has its own mechanism operated by the its collective stock company. On the other hand, it provides half of the urban housing, accommodating thousands and millions of people in the city. With the development of the economy, urban villages are of assistance rather than burdens to the city. Urban villages and the city form a symbiotic circle. Such transformation exerts great influence on urban village renewal, changing the current method of demolishing to refurbishment, coping with the development of the surroundings.

1 城中村的城市化

由综合整治方式主导的城市更新研究

THE URBANIZATION OF URBAN VILLAGES

THE STUDY OF URBAN VILLAGES RENOVATION AS A TYPE OF URBAN RENEWAL

1.1

村城共生：深圳城中村改造研究

SYMBIOSIS IN VILLAGES AND CITIES—RESEARCH ON THE TRANSFORMATION OF SHENZHEN CITY VILLAGES

Yang Zhenyuan 杨镇源
Hu Ping 胡平
Liu Zhenxin 刘真鑫

摘要

深圳历经 40 年的快速发展，蜕变为国际化大都市，高速发展模式下形成城村二元化的城市格局。本文主要剖析城市更新的多种途径，并比较各种实施方式的得失。鉴于无序建设直接导致相关的城市配套无法协调，城中村内部市政设施缺乏合理组织以及城市快速发展所遗留的问题，对城中村环境改造不断探索，形成三种城中村改造类型：综合整治类城市更新、功能改变类城市更新以及拆除重建类城市更新。面对拆除重建类更新的复杂性与矛盾性，以及功能改变对城市所产生的消极影响，城市更新的重心逐渐转向综合整治的发展方向。

关键词

城中村、旧村改造、城市更新、综合整治

1. 城村二元化的城市格局

1978 年中国第一个经济特区成立——深圳，经历 40 年快速发展，从人口不过几万人的一座小渔村，发展成为超千万人口的特大城市。渔村周边建筑日渐增多，生活方式日趋繁复。伴随城市发展开始包围自然村庄，农民转身成为房东，房屋租赁成为主要经济来源，城中村事业由此诞生。根据深圳市规划和国土资源委员会调研数据显示，深圳市城中村用地总规模约 320km^2，约占深圳土地总面积的 16.7%。根据深圳市住建局的摸底调查，深圳共有以行政村为单位的城中村 336 个（其中特区内行政村 91 个），自然村 1044 个（一个或多个自然村组成一个行政村），城中村农民房或私人自建房超过 35 万栋，总建筑面积高达 1.2 亿 m^2，占全市住房总量的 49%。城中村住房与商品房平分住房市场，成为深圳人主要的居住形式之一。由于大部分城中村住宅属于违规建设，建筑环境、物业管理水平等方面限制，促使城中村的房租相对低廉，但恰好为初来深圳打工者提供了最为便捷、经济的居住场所。根据深圳链家研究院的长期监测，实际租住城中村的比例可能达租赁总人口的 60%~70%，居住约有 1200 万人。

ABSTRACT

With 40 years' rapid development, Shenzhen has become an international metropolis. Due to the high-speed development, it has formed a dualistic urban pattern. This paper focus on analyzation of various ways in urban renewal and comparism among different scenarios. Because of guideless development, the related urban infrastructure cannot coordinate effectively. The renewal methods of urban villages are continuously explored, which leads to three types of renewal methods: comprehensive improvement, function alternation, and reconstruction after demolishment. Facing with the complexity and contradiction of large-scale demolition, the negative impact of such operation makes the comprehensive improvement of urban renewal gradually become the preferred mode.

KEY WORDS

Urban villages, old village renovation, urban renewal, comprehensive rectification

1.The dualistic pattern of city and village

In 1978, Shenzhen was established as China's first special economic zone . With 40 years of rapid development, it has raised from a small fishing village with a population of ten of thousands to a megacity with a population of over 10 million. The surrounding buildings in the fishing village are increasing and life is becoming more and more complicated. As the urban development came near, the peasants in local village turned into landlords. House leasing became their main source of income, from which the city-village business was born. According to the survey of Shenzhen Planning and Land Resources Committee, the total land area of urban villages in Shenzhen is about 320 square kilometers, accounting for 16.7% of the total land area in Shenzhen . Another survey conducted by the Shenzhen Municipal Housing and Construction Bureau suggests that there are 336 urban villages in the administrative villages (including 91 administrative villages in the special zone) and 1,044 natural villages (one or more natural villages form an administrative village). There are more than 350,000 houses or private self-built houses, with a total construction area of 120 million square meters, taking up 49% of the city's housing. Urban village housing shares the market with commercial housing and has become one of the main forms of residence in Shenzhen. As most of the urban village houses are illegal constructions and owing to the limits of construction environment, property management level and other aspects, the rent in the village is relatively low, nevertheless it provides the most convenient and economical place for new migrant workers. The long-term monitoring of the Shenzhen Chain Family Research Institute shows that tenants in urban village may represent a percentage between 60% to 70% of the total rental population, with about 12 million residents.

2. 城中村现状环境

城中村的村民私宅往往以个体利益最大化的方式建设，建筑基本占满划分的宅基地。宅基地之间利用留下的间距形成巷道，横纵方向巷道的宽度基本一致，为 2 ~ 3m 左右，仅用于通风不能走车。城中村的建筑群落整体形成了棋盘式路网结构，建筑密度较大，平均都在 35% ~50%（表 1)。

深圳市城中村建筑密度表　表 1
Building Density of Urban Villages in shenzhen　Table1

深圳市城中村建筑密度表（数据来源：2015 年深圳市建筑普查数据）					
城中村	建筑基底面积	总建筑面积	城中村用地范围	建筑密度	容积率
大磡村	44.42 万 m^2	136.83 万 m^2	106.7hm^2	41.60%	1.28
和磡村	5.71 万 m^2	44.37 万 m^2	14.6hm^2	39.10%	3.04
景乐村	1.46 万 m^2	9.39 万 m^2	3.94hm^2	37.10%	2.38
坪山村	6.72 万 m^2	37.00 万 m^2	13.46hm^2	49.90%	2.75

村内普遍以一到两条贯通全村的主街作为主要交通，垂直于主街的为次要交通，主、次街交互串联巷道，形成鱼骨状路网结构（图1）。如景乐新村：以一条市政道路将村落划分为南北两区，每个区内均有巷道连接市政道路，且巷道的宽度基本一致，没有主要道路和次要道路之分。

由于小汽车在早年的城中村建设时期不属于普遍的交通工具，村中基本没有建设集中的停车场地，主街和次街同时兼顾路边停车的功能，剩余道路宽度约为 5m，勉强能够容纳 1 ~ 2 俩小汽车相向而行（图 2, 图 3, 图 4）。

由于城中村建筑密度大，村中用地基本被建筑占满，用作户外活动的公共空间普遍缺乏，仅有的空地多数是在村出入口广场和村内由于某种原因没有建设起来的荒地（图 5，图 6）。而村内公共活动空间受限，街巷道路被车辆占用，村中荒地成为仅有的公共空间但却又没有被合理利用（图 7）。

2. Current situation of urban villages

As the private houses of villagers are often built in the way of maximizing individual interests, the residence footprint of house often takes up the site fully. The space between residence footprint is used as alleys, with a width of 2-3 meters, which is enough for ventilation but limited for traffic. The roads of urban village is in a checkerboard pattern, and the building density is high, with an average of 35%-50% (Table 1).

In the village, streets running through the village are functioned as main traffic roads, and roads perpendicular to them are secondary. The main and the secondary, together with the alleys running across, form a fishbone pattern (Fig. 1). In the case of Jingle New Village: the village is divided into north and south two parts by a city road. Each part is connected to the city road by alleys basically with the same width, with no hierarchy.

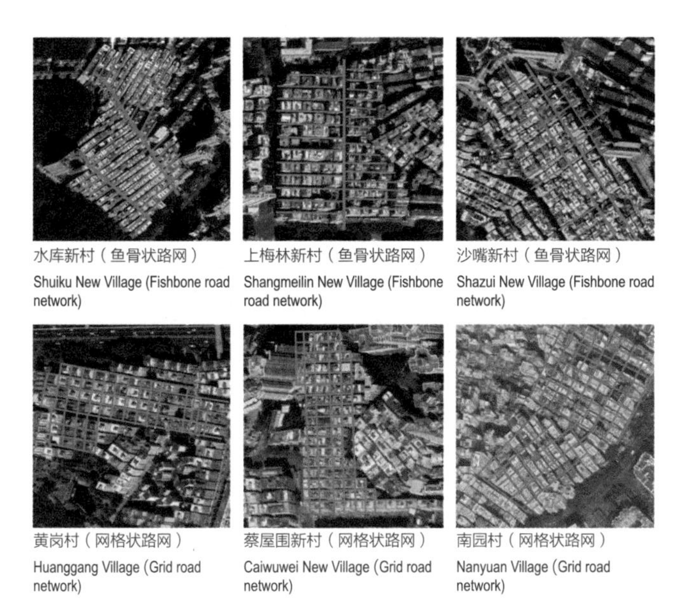

水库新村（鱼骨状路网）
Shuiku New Village (Fishbone road network)

上梅林新村（鱼骨状路网）
Shangmeilin New Village (Fishbone road network)

沙嘴新村（鱼骨状路网）
Shazui New Village (Fishbone road network)

黄岗村（网格状路网）
Huanggang Village (Grid road network)

蔡屋围新村（网格状路网）
Caiwuwei New Village (Grid road network)

南园村（网格状路网）
Nanyuan Village (Grid road network)

图 1 城中村道路路网结构
Fig.1 City Village Road Network Structure

Since small vehicles were not the main transportation for residents in the early stage of urban village construction, there was barely any parking lot. On-street parking is provided by the main street and secondary street. Space left by parking vehicles on streets was about 5 meters, which is barely enough for one or two cars to move in opposite directions (Fig.2, Fig.3, Fig.4).

无序规划直接导致相关的城市配套无法协调，城中村内部市政设施缺乏科学组织。电力通信电缆基本暴露在外，电线缠绕、路线老化、线路私接现象严重，具有较大的安全隐患（图 8)。

给水、排水管道残缺，排水不及导致内涝现象时有发生。生活污水、雨水尚无分流排放，给村外河道的治理造成了很大困难（图 9)。密集的居住人口，以及整体的人口素质偏低导致公共空间卫生环境较差，包括垃圾收集点缺乏管理，垃圾收集转运系统不完善，垃圾桶（车）肮脏简陋，垃圾乱扔成堆等相关现象的产生（图 10）。

图 2 和磡村内道路
Fig.2 Hekan Village Road

图 3 上梅林新村内道路
Fig.3 Shangmeilin Village Road

图 4 蔡围屋八 · 九坊村内道路
Fig.4 Caiweiwu Village Road

图 5 和磡村内荒地
Fig.5 Harmony in Hekan Village

图 6 麻磡村内荒地
Fig.6 Harmony in Makan Village

图 7 和磡村内荒废公共空间
Fig.7 Abandoned public space in Hekan Village

图 8 和磡村内电线暴露杂乱
Fig.8 Harmony in Hekan Village

图 9 大磡村内河道环境较差
Fig.9 Harmony in Makan Village

图 10 大磡村内垃圾桶肮脏
Fig.10 Public Space Not Properly Utilized in Hekan Village

3. 城中村环境改造的基本思路

我国经历了快速的城市发展，2017 年中国城市化率提高至 58.52%，传统的增量发展逐渐转换为存量发展，如今的城市更新需要随着时代同步发展。[1] 截至 2005 年，深圳的可建设用地已不足 200km^2。[2] 现在深圳市的城市化程度已达 100%，近几年的土地供应十分紧张，目前获得新增用地只有两种方式——旧城改造和填海。这在某种程度上对深圳市的持续发展构成一定阻力。相比填海造城的漫长周期和远离城市的地理位置，旧城改造的方式是解决眼前城市土地供应紧张的捷径，并且能够充分提升靠近城市中心的土地价值。因此，旧城改造成为近几年深圳城市发展新项目的主导方式。自 2009 年 12 月 1 日起，深圳市政府开始实施《深圳市城市更新办法》对旧

Due to the high density of buildings in the village, with land occupied by buildings, there is scarcely public space for outdoor activities. The only open space is the entrance plaza and wastelands that have not been built(Fig.5, Fig.6). However, the public activity space in the village is limited, the roads in the street are occupied by vehicles, so the undeveloped-land in the village becomes the only public space but it has not been rationally utilized (Fig.7).

Disordered planning directly leads to the inability of coordination of relevant urban facilities, and the lack of scientific organization of municipal facilities in urban villages. The power communication cable is basically exposed,The problems of wire winding, route aging and private connection of cable lines are serious, which have great potential safety hazards (Fig.8).

The water supply and drainage system is incomplete, and the drainage is not enough, causing internal defects. There is no diversion of domestic sewage and rainwater, which has caused great difficulties in the management of rivers outside the village (Fig.9). The density of population and the low quality of the population lead to poor public health environment, including lack of management of garbage collection points, imperfect garbage collection and transportation systems, dirty trash cans, and littering behaviour. (Fig.10).

3. Basic ideas for environmental transformation in urban villages

Through rapid urban development, In 2017, China's urbanization rate increased by 58.52%. The traditional incremental development gradually turned into stock development. Today's urban renewal needs up-to-date development. [1] By 2005, the construction land in Shenzhen has been less than 200 square kilometers. [2] The urbanization degree of Shenzhen has now reached 100%. In recent years, land supply is very tight in Shenzhen. At present, there are only two ways to obtain new land: old city renovation and reclamation. This has certain resistance to the continued development of Shenzhen. Compared to the relatively long cycle of reclamation and the geographical location away from the city, the old city transformation method is to solve the short-term shortage of urban land supply and to fully enhance the value of land close to the city center. Therefore, the transformation of the old city has become the leading mode of new urban development projects in Shenzhen in recent years. Since December 1, 2009, the Shenzhen Municipal Government has implemented the “Shenzhen City Renewal Measures” to guide the transformation. As it is implied, there are three types of urban village renewal: comprehensive remediation (Fig.14), functional change (Fig.15) and

城改造的方式进行了指引。其中，该文件对于城中村改造的办法划分成三种类型：综合整治类城市更新（图 14）、功能改变类城市更新（图 15）和拆除重建类城市更新（图 16）。对于区位较好、临近轨道站点、容积率较低的城中村以拆除重建的方式进行整治。拆除重建类城市更新能够有效整合土地资源，梳理片区的路网结构、改善交通，同时政府采取收回一部分国有的土地用于布置公共配套设置如学校、医院等，充分提高土地利用效率。对于容积率较高、拆除难度较大的城中村采取改造提升的方式进行综合整治。而功能改变类城市更新更适用于厂房类建筑，因为厂房建筑灵活的结构能适用不同类型的城市功能。

1）拆除重建类更新的复杂性与矛盾性

在这三种城市更新的办法中，由于拆除重建类的项目使得短期利益达到最大化，即便推进项目难度较大，也使其成为开发商和村民最积极参与的更新方式。以位于南山区科技园的大冲村为例，早在 2002 年，大冲村旧改就已经被深圳市政府列为旧村改造项目的重要试点村之一。原村民约 1007 人，居住人口达 6 万余人。更新改造前的大冲村以居住用地、工业用地及园地为主，有各类建筑 1400 多栋，总建筑面积 106 万 m^2，其中原农村集体物业约 40.4 万 m^2，382 户原村民拥有的私宅约 69.4 万 m^2，是原深圳经济特区内规模较大的城中村。[3] 但在拆迁和补偿问题的困扰下，项目进展受阻。直到 2009 年才得以进一步发展，7 年的时间跨度也为建设本身提出诸多考验，在政府的强力推动下，大冲股份公司才与华润集团正式签订了《大冲旧改合作意向书》，此时该项目算是迈出了实质性的第一步。大冲村作为全市最大的城市更新项目，也是南山区旧改重点推进工程，整体改造面积达 68.5hm^2（图 11），改造后的大冲村项目开发建筑量约为现状建筑量的 2.5 倍，容积率达 5.74，其中居住用地最大容积率为 8.12（图 12）。规划中除保留现状的大王古庙、郑氏宗祠以及几棵古树外，对规划范围内的其他用地进行了重新整合，拆除重建之后的大冲村拥有新的规划理念。作为深圳市高新技术产业园区的配套基地，本区的功能配置除一般的居住、公寓外，更增加了相关的商业、文化、娱乐及专业展场等多项功能，如深圳万象天地的商业建设面积达到 78 万 m^2，使大冲村成为功能齐备的生活配套区，增加高新产业的吸引力，并以此带动大冲村的城市化改造进程（图 13）。

图 11 改造后大冲村：华润万象天地

Fig.11 After the renovation, Dachong Village: China Resources Vientiane World

图 12 改造前大冲村谷歌卫星地图

Fig.12 Before the Transformation, Dachong Village Google Satellite Map

图 13 改造后大冲村谷歌卫星地图

Fig.13 After theRenovation, Dachong Village Google Satellite Map

demolition reconstruction (Fig.16). For the villages with good location, closed to subway station and of low floor area ratio, the villages will be rehabilitated by demolition and reconstruction. Demolition and reconstruction can effectively integrate land resources, sort out the road network of the district, and improve traffic condition. At the same time, the government will take back some of the state-owned land for public facilities such as schools and hospitals, raising the land use rate. For the urban villages with high plot ratio and difficult to be removed, they will be upgraded and improved. Functional change is more suitable for industrial buildings, because the flexible structure of the building can be applied to different types of urban functions.

图 14 综合整治：水围柠盟公寓

Fig.14 Comprehensive rectification: Shuiwei LM Apartment

图 15 功能改变：华侨创意文化园

Fig.15 Functional Change Category: OCTLOFT

图 16 拆除重建：华润万象天地

Fig.16 Demolition and Reconstruction: China Resources Vientiane World

1）The complexity and contradiction of demolition and reconstruction

Among the three urban renewal methods, demolition and reconstruction is prefered by developers and villagers, for it maximizes the short-term benefits, despite having difficulties in promoting. Take Dachong Village,in Nanshan District Science and Technology Park as an example. As early as 2002, Dachong Village was one of the important pilot villages of the old village renewal project by the Shenzhen Municipal Government. The original villagers were about 1,007, with a population of more than 60,000. Before the renovation, Dachong Village is mainly composed of residential land, industrial land and gardens. There are more than 1,400 buildings of various types, with a total construction area of 1.06 million square meters, of which the original rural collective property is about 404,000 square meters, and 382 houses owned by the original villagers. It is about 69,4000m^2 which is a large-scale urban village in the former Shenzhen Special Economic Zone.[3] However, with the trouble in compensation, the progress of the project was hindered. It was not until 2009 that it was further promoted. During these years, new challenges were also posed. Under the strong promotion of the government, Dachen Co., Ltd. officially signed the "Dachong Old Reform Cooperation Letter of Intent" with China Resources Group. The project is the first step. As the largest urban renewal project in the city, Dachong Village is also the key promotion for the renovation of Nanshan District. The overall renovating area of Dachong is 68.5 hectares (Fig.11). The reconstructed Dachong Village project has a construction capacity of about 2.5 times the current construction volume and a plot ratio of 5.74, of which the maximum plot ratio of residential land is 8.12 (Fig.12). In addition to retaining the status quo of Dawang Ancient Temple, Zheng's Ancestral Hall and several old trees, other land in the planning area was re-integrated. After demolish and reconstruction, Dachong village was defined as a supporting base of Shenzhen High-tech Industrial Park, with additional commercial, entertaining, cultral and exhibiting functions to residence.For example, the commercial construction area of Shenzhen Vientiane World has reached 780,000 square meters, making Dachong Village

虽然拆除重建类的城市更新项目带给了城市优越的配套设施，但是也带来了一系列的社会问题。深圳的高房价使得外来人口的生存空间受到挤压，城中村低廉生活成本一直为深漂的人们提供庇护。如果说深圳是一座圆梦之城，那城中村就是巨大梦想的孵化器。虽然城中村生活环境脏乱，并且存在安全隐患，但是城市的包容意象却实实在在为底层的人员提供生活空间。拆除重建后的城中村虽然以丰富的空间活力屹立在城市之中，但是高额的房租成本却不断地挤压外来租户的生活空间，诸如白石洲村的旧改项目，导致其周围的房租呈跳跃式增长。2017 年街对面的侨城豪苑一房一厅（建筑套内面积约 40m^2）月租金已经到达 4500 元（每平方米平均月租金超过 100 元）[4]，不断攀升的房价使得外来租户不得不再一次进行迁徙。城中村原有密集的充满活力的街巷、形式百态的商业、狭窄巷道里穿梭的身影连同城市的记忆也消失不在，同时拆除重建过程中迁移走的 15 万租客在忙乱的城市中无处安身，这是拆除重建类城市更新所付出的沉重代价，变相将低收入人群从城市驱赶到市郊。

2）功能改变类城市更新的机遇与启发

地处深圳另一处的城中村缺少都市核心区的绝佳区位，但由于地缘的优势，村民选择了对自己的物业进行自发性的功能改造来增加收入，原本出租屋的主导功能附带了经营性、服务性的商业功能。村民的自发性改造促进了本土性的私营产业，特色性的城中村街区化城市空间由此诞生。位于深圳东部沿海的较场尾片区曾经是滨海渔村，随着人均可支配收入提高、旅游产业渐渐兴旺，2007 年较场尾改造的第一家民宿开业。2011 年深圳市举办大运会为民宿产业后续发扬光大提供了重要契机，政府划拨了一笔专项资金，用于完善较场尾的基础设施。[5] 随着环境质量提高，人口密度增大，民宿的事业日渐兴旺。至 2012 年底，已有逾百家民宿开业，村民也开拓了快速发展经济的道路（图 17，图 18）。利好的资源价值同样吸引到社会资本，2013 年曾经有开发商尝试推动拆除重建式的更新改造，但在约三分之一的业主的坚持下，开发计划未能实施。改造整治仍作为城市更新模式，在得到政府认可的同时，加大投入进行基础设施的改造。大量用于民宿经营的村屋，导致居住人口大幅增长，原有村中的供电、供水和排污承载不了民宿酒店用量，而公共设施的提升恰到好处地支持

图 17 较场尾卫星地图
Fig.17 Jiaochangwei Satellite Map

图 18 较场尾民宿
Fig.18 Jiaochangwei Homestay

a fully functional living community, increasing the attractiveness of high-tech industries, and driving the urbanization process of Dachong Village (Fig.13).

Although the demolition and reconstruction of urban renewal projects has brought the city's superior supporting facilities, it also brings a series of social problems. The high housing prices in Shenzhen have squeezed the living space of the migrant population, while the low cost of living in the urban villages provides shelter for the migrants. In Shenzhen, the urban village is a dream incubator. Despite the messy environment and potencial safety hazard, it implies the tolerance of the city and provides living space for the underclass. Although the demolition and reconstruction of the village provides abundant space, the high rent constantly eliminates affordable housing for the migrants. For example, renovation project of Baishizhou Village led to a rapid growth in rental in surrounding estates. The rent of the one-bedroom and one-bedroom hall of the Overseas Chinese Town (the building covers an area of about 40 square meters) across the street in 2017 has reached 4,500 yuan (the average monthly rent per square meter is more than 100 yuan) [4]. Rising continuously ,the high rental has forced foreign tenants to migrate again. In urban village, intensive and vibrant streets, the form of business, and the narrow lanes disappeared with the memory of the city. At the same time, the 150,000 tenants who migrated during the reconstruction process had nowhere to live in the busy city. This was a heavy price for the demolition and reconstruction of urban renewal, and the low-income group was driven from the city to the suburbs in disguise.

2) Opportunities and inspirations for functional change city renewal

The city center located in another part of Shenzhen lacks the excellent location of the urban core area. However, due to the geographical advantages, the villagers chose to carry out spontaneous functional transformation of their own properties to increase their income. The original leading function of rental housing became a service-oriented commercial business function. The spontaneous transformation of the villagers promoted the local private industry, and the characteristic urban village-street urbanization space was born. The Changchang tailings area on the eastern coast of Shenzhen used to be a coastal fishing village. With the in per capita disposable income and the tourism industry increasing gradually, in 2007, the first B&B was opened, and the Universiade held in Shenzhen in 2011 provided an important opportunity

了改造的事业。海边生态较敏感和脆弱，基础设施的完善提升既保障了良性发展的整治模式也保护了自然生态不受侵害。

今日的较场尾悄然成为节假日旅游胜地。自发性的改造在后续获得地方政府的支持至关重要，正因如此村内的基础设施才得以完善，以适应居住人口和建筑功能改变所带来的消耗增长。另一方面自发式改造也存在一些问题，目前在较场尾经营民宿的商家绝大多数从村民手上以租赁的方式与其合作经营，而村民作为房东收租。由于租期以6年为主，某种程度上也限制了商家对建筑环境大刀阔斧的投入改造。由于村民招商的工作不设门槛，虽然一房难求，但是商家水平的参差不齐也使得片区的整体氛围和环境管理存在问题。现有的500栋单体建筑，许多建筑的现状不佳，建筑改造并不精美，未能凸显滨海风情地域特色，其空间及外观均有进一步改善的可能。[5]

3）综合整治类城市更新的尝试

在近年来的实践中，城市更新的推动难度较大，牵涉的利益主体较多，更新过程中导致原村民暴富，租客不断迁徙，中低收入人群无房可租等社会问题不断产生，而不拆式改造的城中村综合整治便成为城市更新的另一选择。对于合法用地占比低，容积率过高，拆除重建可操作性低，甚至不具备拆除重建条件的城中村，综合整治也是提升片区环境的最佳选择。综合整理保留原有城中村建筑，不涉及土地及物业产权调整，在原有建筑的基础上，对城中村老旧的基础设施进行提升（图19），包括交通的梳理（图20）、公共空间与设施（图21）、卫生条件、消防安全、环境风貌改善等方面的问题，对保留的原有建筑进行改造（图22）、功能升级（图23），以使城中村的形象和功能得到全面提升。 城中村的综合整治不仅是对城中村改造的环境提升，还是对现有城市空间和建筑功能的重塑与完善，促进城中村与周边城市环境相互融入，互补共存。

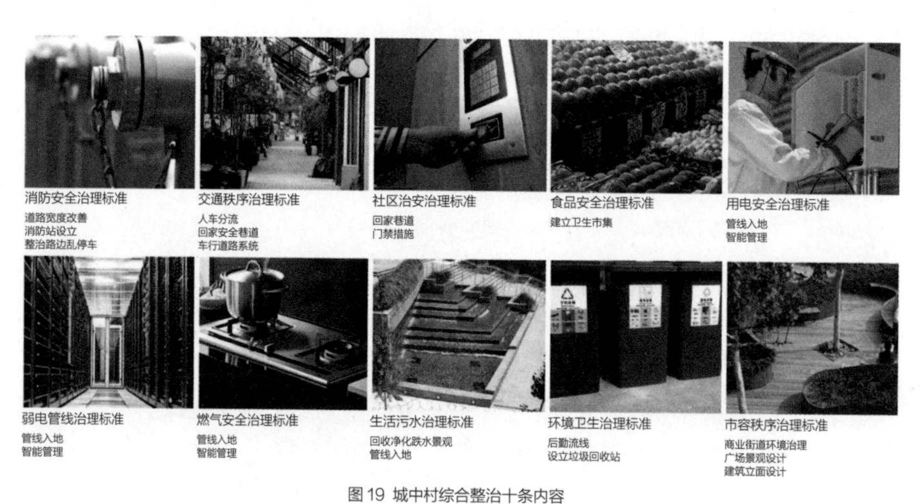

图19 城中村综合整治十条内容

Fig.19 Ten Elements of Comprehensive Rectification in Urban Village

for the follow-up development of the residential industry. The government allocated a special fund to improve the infrastructure at the end of the field. [5] With the improvement of environmental quality and population density, the business of the hotel increased gradually. By the end of 2012, more than 100 B&B had been opened, and villagers had found a way to rapidly develop their economies (Fig.17, Fig.18). The value of good resources also attracts social capital. In 2013, developers tried to promote the demolition and reconstruction, but, with the insistence of about one-third of the owners, the development plan was not implemented.Though, reconstruction and rectification is still regarded as an urban renewal model, and it is recognized by the government and increased investment in infrastructure transformation. The village houses used for the operation of the property house, with the substantial increase of the resident population, the power supply, water supply and sewage in the original village can not bear the large amount of hotels. The improvement of public facilities supports the transformation.The improvement of the infrastructure not only protects the remediation model of benign development but also protects the natural ecology, sensitive and fragile, from infringement.

Today's JiaoChangwei village has become a holiday destination. Spontaneous transformation is crucial to the subsequent support by local governments, promoting the improvement of infrastructure in the village, accommodating the growth of consumption resulted from changes in the population and building functions. Meanwhile, there are some problems in spontaneous renewal. At present, most of the businesses in the JiaoChangwei village are in buildings rented from the villagers. However, the lease period is mainly 6 years,which limits the transformation of the building environment to a certain extent. Since the villagers' investment work does not have a threshold, the uneven level of business has also caused problems in environmental management of the district. Many of the existing buildings are in poor condition, and the architectural renovation is not exquisite, failing to highlight the coastal style. Their space and elevation can be further improved. [5]

3) Urban refurbishment

In recent years, as more stakeholders involved, tenants evicted tenants evicted with fewer low-income housing, compared to demolishing, city refurbishment is a preferable mode of renewal. It is more suitable also for its ability to improve the environment of urban villages with low proportion of legitimate land, high volume ratio, and low operability of demolition and reconstruction. In this way, the original buildings in urban village are preserved and old infrastructure is upgraded based on the original constructions preserving the original tenure security (Fig.19). The process includes traffic carding(Fig.20), improvement of facilities(Fig.21), public area and amenities,

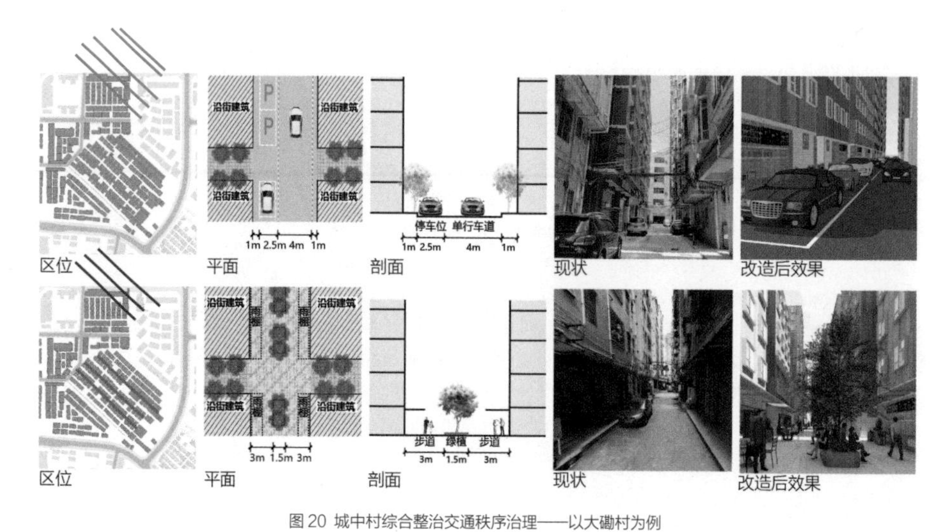

图 20 城中村综合整治交通秩序治理——以大磡村为例

Fig.20 Comprehensive Improvement of Traffic Order Management in Urban Villages:A Case Study of Dakan Village

图 21 城中村综合整治公共空间改造——以和磡村为例

Fig.21 Renovation of Public Space in Comprehensive Renovation of Urban Villages: A Case Study of Hekan Village

图 22 城中村综合整治建筑立面改造策略——以和磡村为例

Fig.22 Strategy for Renovation of Building Facade in Urban Village: A Case Study of Hekan Village

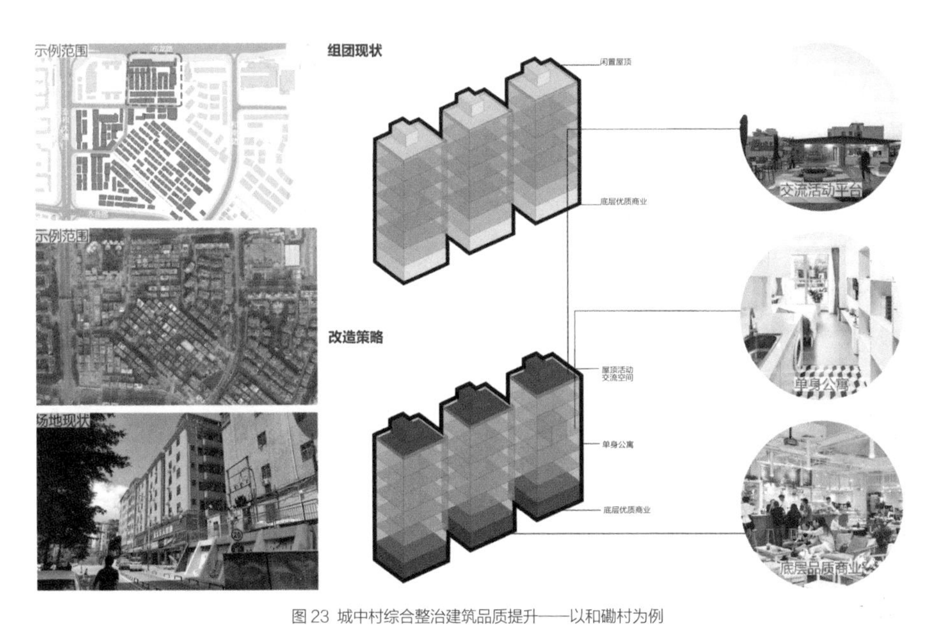

图 23 城中村综合整治建筑品质提升——以和磡村为例

Fig.23 Comprehensive Improvement of Building Quality in Chengzhong Village： A Case Study of Hekan Villagee

对于城市中心区的城中村即便是没有形成自发性的升级，但由于其地理位置优越便利具备较高的土地价值，促成了政府推动企业作为主导的改造方式。以福田区水围村的综合整治为例，推动政府、企业、村民共同合作，采用取长补短的合作改造模式。水围村规划面积约 8000m^2，共 35 栋统建农民楼，其中的 29 栋改造为 504 间人才公寓。[6] 政府负责公共配套与市政设施、企业负责项目改造与运营、村集体股份公司发挥基层协调和配合。公共配套部分，由区政府投资，对管道燃气、给排水管网、供电系统等配套设施进行综合整治。针对村民楼部分的整理，深业集团作为公寓改造和运营方，由其向水围股份公司统租 29 栋村民楼，按照人才住房标准改造后出租给区政府，区政府再以优惠租金配租给辖区产业人才，而补贴的金额就相当于政府整治城中村的成本。在调研过程中，咨询水围公寓的工作人员得知，水围公寓的租金中由政府提供的补贴达到 75 元 /m^2，改造成公寓的总建筑面积 1.3 万 m^2，一年下来仅水围人才公寓的租金政府就需要补贴 1100 万元左右。该项目也成为深圳市首个利用城中村居民楼改造成人才保障性住房的试点项目，既盘活了老旧城中村社区，又为城市中心提供了保障性住房，从而形成多方共赢的局面。

改造后的街区仍保持原有的城市肌理与城中村的空间尺度，改造对建筑功能与流线进行了调整。建筑 1、2 层的功能改造为一拖二形式的底层商业，由村集体股份公司统一招租运营。在引入商户的同时，也为该区的住户提供餐饮、娱乐、休闲等服务设施。3 层及以上部分作为人才保障性住房和租赁公寓。建筑之间的巷道宽 2.5 ~ 4m。[6] 将稍宽的东西向巷道改造成步行商业街（图 24），其余较窄的巷道作为居民出行的通道，私密性也较好，并在其中加建电梯，围合成入户前院子（图 25），提高了居民上下出行的便捷性与私密性。巷道上空三层和五层的位置均增设空中连廊（图 26），把楼栋联系起来，提高了电梯的使用效率，不仅方便住户更为便捷地抵达各个功能空间，也成为居民休息、交流的公共活动场所，营造了立体式生活街区。

作为综合治理的成功案例，水围柠檬人才公寓拥有其独特的模式（图 27）。首先，它是以政府为主导的城市更新整治项目，政府出资将城中村整租下来进行改造，并以低价租给辖区内产业人才，这为人才保障性住房的建设提供了一个很好的借鉴方向。相比于拆除重建

图 24 水围村内步行商业街
Fig.24 Shuiwei Village Walking Commercial Street

图 25 水围公寓电梯院子
Fig.25 Shuiwei Apartment Elevator Yard

图 26 水围公寓空中连廊
Fig.26 Shuiwei Apartment Aerial Gallery

sanitary and fire-safety conditions.etc, promoting a new image[fig.22] and fulfilling the requirement for functions[fig.23]. Despite improvement of the village, the renewal enhances the functions and the overall environment of the city, blurring the boundary between the urban village and the city, making a better coexistence.

For the urban village in centre of the city without spontaneous upgrading, though, its comparetively higher land value due to geographical location makes the government-led promotion the leading renewal method. Take the renovation of Shuiwei Village in Futian District as an example, it promotes cooperation among the government, enterprises and villagers, and adopts a cooperative transformation mode that complements each other. The planned area of Shuiwei Village is about 8,000 square meters. A total of 35 buildings are built, of which 29 are transformed into 504 affordable apartments for the young professionals. [6] The government is responsible for public facilities and municipal facilities, and the enterprise is responsible for project transformation and operation, and the village collective stock company plays the role of coordinator and cooperator.The public supporting part will be invested by the district government to update the supporting facilities such as pipeline gas, water supply and drainage pipe network and power supply system. Shum Yip Group Limited, as the invester and operater, arranged a unified rental of 29 buildings from Shuiwei Co. Ltd., then leased them to the Futian Government after renewing them to the standard of affordable housing to the young professionals. The district government then rented it at a preferable price. For the industrial talents in the jurisdiction, the amount of subsidies is equivalent to the cost of the government to rectify the villages in the city. In the investigation, the rent of the Shuiwei Apartment ,subsidized by the government ,was 75 yuan per square meter, and the government needs to subsidize about 11 million yuan per year for the total renovation area of 13,000 square meters. The project has also become the first pilot project in Shenzhen to transform the residential buildings in the urban villages into affordable housing for the talents. Renewing the old village and at the same time easing housing burden for at the center of the city, the project formed a multi-win situation.

The reconstructed block still maintains the original urban texture and the spatial scale of the urban villlage, and adjusts the building function and flowlines. The functions of the first and second floors of the building are transformed into commercial area, which is uniformly leased and operated

的城市更新，城中村综合整治方式是更为柔性的改造方式。首先拆除重建的综合成本非常高昂，每当拆掉一座城中村，利益就流向原住民、开发商手中，而租户成了最大的受损方，他们必须要搬去更偏远的地方，这对构建多元、包容且充满活力的城市中心不利。而综合整治改造的成本相对较低，不但可以保留城市的历史与文脉，原租户也依然有机会居住在城市中心，对构建和谐社会起促进作用。不过实施“水围模式”的综合整治有一定的难度，该模式需要借助政府出资将改造好的城中村整租下来再以低价出租。而通过协调整村统租来实施整体改造就导致该村全部物业长达半年以上的空租期，自然会增加改造成本。如果不考虑政府补贴的人才公寓租金，改造后的长租公寓即便租金有所上涨也非常畅销。

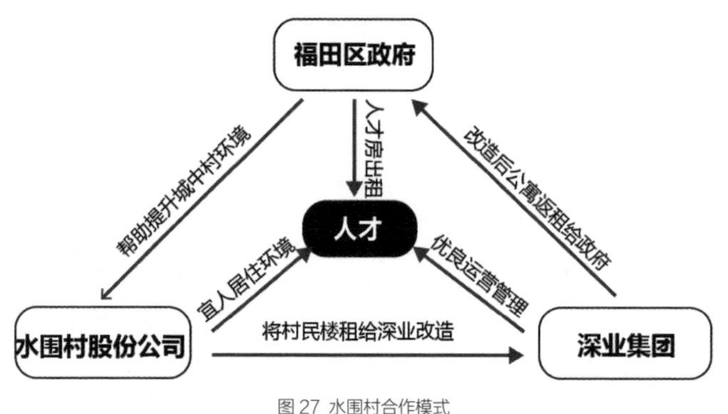

图 27 水围村合作模式

Fig.27 Shuiwei Village Cooperation Mode

图 28 景乐南 1 号楼改造前

Fig.28 Before the renovation of Jingle South Building No. 1

4. 城中村改造市场化的成效

由于深圳市全域的城中村规模巨大，其所隐含的物业价值也被不同的地产商看中。其中社会影响力更大的综合整治项目要数万科推出的万村计划。万科的万村计划起源于 2017 年 7 月，深圳市万村发展有限公司在万村大梅沙总部成立；同年 8 月，“万村复苏”的首个试点——岗头新围仔村启动改造；12 月，新围仔村首栋长租公寓样板楼对外公布。岗头新围仔村位于龙岗坂田，靠近华为、富士康等知名科技企业园区。周边临近地铁，坐 10 分钟公交车可到 5 号线地铁站，租户多是周边大企业员工。新围仔村共有 226 栋房屋，万科目前已与村民签约 20 余栋，并将在这里改造出 1000 多套租赁公寓产品。现在有两栋已改造完成，2018 年 3 月首批推出房源 108 套，共 4 种户型，面积约为 18 ~ 22m^2，未改造前单房均价为 800 元，一房一卫价格区间为 1100 ~ 1200 元，两房一卫均价为 1250 元，改造后的泊寓（万科长租公寓产品）价格区间为 798 ~ 1398 元（含家私家电），租金涨幅控制在 10%左右，居住品质也随之有了较大提升。

图 29 景乐南 1 号楼改造后：青年公寓

Fig.29 After the renovation of Jingle South Building 1: Youth Apartment

万村计划陆续在玉田村、景乐新村、上角环村、怀德芳华小区、

by the village collective stock company. At the same time, it also provides catering, entertainment and other service facilities for residents. The third floor and above are used as affordable housing and rental apartments.

Alleys, gaps between the buildings, are 2.5 to 4 meters wide. [6] Alleys directing from east to west, slightly wider, are transformed into a pedestrian commercial street (Fig.24), and the others are used as passages for residents. An elevator is added in the enchance courtyard respectively(Fig. 25) facilitating the vertical transportation and privacy of each block and down and privacy of each block. Sky corridora are constructed between the towers at the three and fifth-stories' height (Fig.26). Thus, the connection between buildings and the efficiency of the elevator are enhanced. Meanwhile, this 3D network forms an important public space for the community, providing extensive space to the individual units.

As a successful example of urban village refurbishment, Shuiwei Ningmeng Apartment Project has its own unique mode (Fig.27). First of all, it is a government-led urban renovation project. The government rented the whole village for renovation, then rent it to the industrial talents within its jurisdiction at a low price. It provides a possible way for the construction of talent-guaranteed housing. Compared with the urban renewal of demolition and reconstruction, it is more flexible. First, the cost of demolition and reconstruction is very high. Whenever a village in a city is demolished, the benefits flow to the locals and developers, while the tenants are forced to move to more remote places for low-rental housing. It does harm more than good to construct a diverse, inclusive and vibrant urban centre. The cost of renovation is relatively low. Rather, this transformation mode can preserve the history and context of the city and enable the original tenants to live in the centre of the city, promoting a harmonious society. However, it is difficult to implement the transformation mode of the Shuiwei village. This mode needs funds from government for renovation and low rent. Additionally, coordinating unified renting of the whole village to conduct the overall transformation will result in an empty lease period of more than half a year in the entire property of the village, which will naturally increase the cost of renovation. Moreover, the renovated long-term rented apartment can still be in great demand even without subsidy from the government.

石厦村、平山村、大梅沙村等村落进行综合治理，截止 2018 年中已涉及 80 ~ 90 个城中村（如图 28—图 31）。

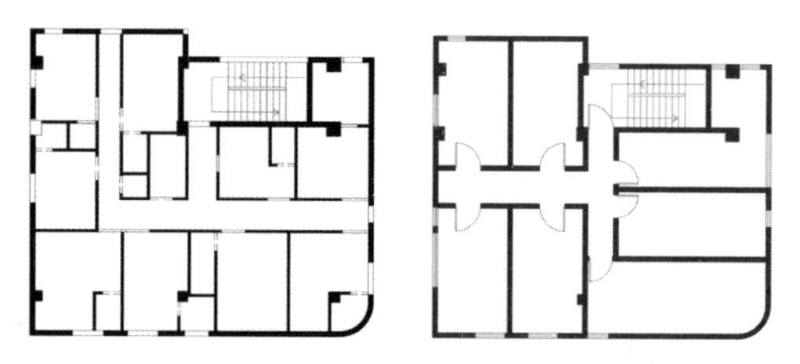

图 30 景乐南 1 号楼改造前平面图

Fig.30 Front view of the renovation of Jingle South Building 1

图 31 景乐南 1 号楼改造后平面图

Fig.31 Plan after renovation of Jingle South Building 1

万村计划的改造模式与水围村统筹改造的方式不同（图 32）。万村计划从村民个体出发，与村民直接签约，通过对每一栋村民自建楼的改造来实现整个村的改造。这种方式可以避免整体统筹的空租期，能进一步压缩控制成本，避免租金大幅度上涨所带来社会影响，但是对企业的商务谈判团队有较高的要求。不足之处就是村内的公共设施并不能与建筑改造一并统筹建设，项目的公共设施如连廊、公共电梯、公共配套功能等会难以落地。相比于政府主导的城中村综合整治，企业资本主导的城中村更新在市场经济的引导下更具有活力，能推行的范围更加广泛。

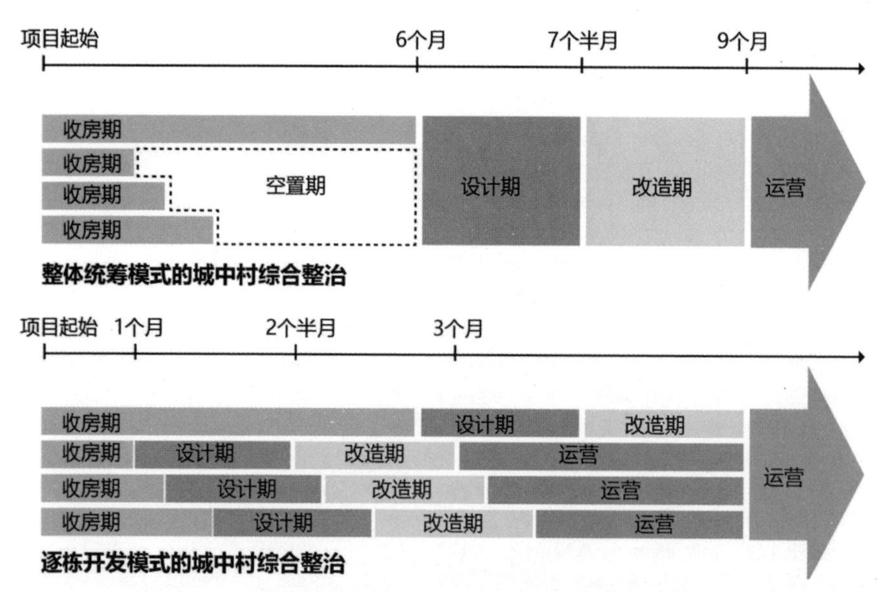

图 32 两种模式下的城中村综合整治时间轴

Fig.32 The comprehensive rectification timeline of Chengzhong Village in two modes

4. The market-oriented renewal of urban villages

The value of the property in urban villages for their large scale is also noticed by different real estate developers. Among current renovation projects, the most influencial one is Wancun Project, which is launched by Vanke.Wancun Project starts from July 2017. Shenzhen Wancun Development company was established in Wancun Dameisha Headquarters. The first project of "Wancun Recovery" – Gangtou Xinweizi Village started renovation in August. In December, the first long-term rent apartment building model in Xinweizi Village was announced. Gangtou Xinweizi Village locates in Bantian,Longgang, beside the well-known technology enterprise parks such as Huawei and Foxconn. The apartment is only 10 minutes away by bus from the subway station of Line 5. Most of its tenants are employees of the surrounding companies. There are 226 buildings in Xinweizi Village. Vanke has signed more than 20 of them with the villagers. Vanke will transform the buildings into more than 1,000 sets of rental apartment. Up to now, two buildings have been rebuilt. In March 2018, the first 108 flats of 4 types were launched. Each flats is about 18-22 square meters. Before the renovation, the average price for a single flat is 800 rmb. For the price of 1100-1200 rmb, the average price of an apartment with two rooms and one bathroom is 1,250 yuan. The price of the remodeled apartment (Vanke long-rent apartment products) is from 798 to 1398 rmb (including furniture appliances), and the increase of rent is controlled at about 10%. Despite the raise in price, the quality of living space is revived.

Wancun Project has been successively carried out in villages such as Yutian Village, Jingle New Village, Shangjiaohuan Village, Huaidefangfang Community, Shixia Village, Pingshan Village and Dameisha Village. By the end of 2018, the number of the villages has reached between 80 and 90. (Fig. 28-Fig.31).

The transformation mode of the Wancun Project is different from Shuiwei Village (Fig.32). The Wancun Project contracts directly with the villagers to realize the transformation of the entire village through the transformation of villagers' own building. This method can avoid the the empty lease period, and further reduce the cost, avoiding the negetive social impact of the rapid increase in rents. However, there are challenges for the business negotiation team of the company. The downside of the method is that the improvement of public facilities, such as corridors, public elevators, and supporting functions cannot coordinate with the building renovation process. Compared with the government-led intergrated renovation of urban villages, the enterprise-led village renewal is more dynamic under the guidance of the market economy, when it is feasible in more cases of renewal.

5. 城中村城市化后续工作

随着我国市场经济的快速发展和住房体制的不断完善，住房租赁市场逐渐发展起来，在城中村的发展中，住房租赁关系中承租人的弱势地位、保护承租人居住利益的重要性已经得到广泛承认，统租市场的规范化将有助于租赁市场的秩序进行。深圳是人口流动率极高的城市，租赁市场的出租屋成为深漂人员的首选之居，有 63.7% 的人通过租赁的方式来获取住房，但是政府所提供的中转房只占租赁市场的 0.8%，由于没有健全的廉租房系统，导致城市人口流动频繁。[7]

统租市场的规范化有助于城市更新的发展，这其中不乏需要政府的支持与法律的保护，在综合治理之初，也可从设计方向入手，集中优化中小户型的供应需求，以使其适应租赁市场变化的需求。政府在推进改造的过程担当着重要的角色，低廉的住房价格与优越的居住环境有助于城市更新发展，其运作模式也有全国推广的价值；相关法律政策的出台有助于市场的维稳与租赁双方合法权益的保护，特别是针对租客的保护，避免租客被业主勒令限期搬出，却没有获得相应的违约补偿。维持租赁市场稳定，对合法住房出租给予一定补贴、权益保护等支持，健全和完善住房租赁市场管理法规，逐步推进住房租赁市场法规的实施等政策措施，大力推进住房租赁市场规范、有序、持续和健康发展。

5. Follow-up work on urbanization of urban villages

With the rapid development of China and the market economy and the continuous improvement of the housing system, the housing leasing market gradually developed. In the development of urban villages, the weak position of tenants in the housing leasing relationship and the importance of the tenant's residential interests have been widely realized. The standardization of the market will improve the order of the rental market. The population mobility rate in Shenzhen is extremely high. The rental housing in the rental market has become the preferred residence for the migrants. 63.7% of them have obtained housing through leasing, with only 0.8% living in interim housing by the government. The incompleteness of the low-rent housing system leads to frequent population mobility. [7]

The standardization of the unified renting market will help the development of urban renewal. While government and the protection of the law are required. The renovation may also starts from small and medium-sized units designed to cope with the market need. In this process, government plays an important role: controlling housing prices which is an effective method that worths publishing in the whole country, and conducting laws and policies which protect the legitimate rights and interests of both parties of the lease-tenants especially, avoiding their eviction by landlords without getting compensation for the breach of contract. The government shall maintain the stability of the rental market, provide certain subsidies for legal housing rental, preserve rights and interests, improve the housing rental market management regulations, and gradually adopt policies and measures such as housing rental market regulations, and vigorously promote a regulated, ordered, sustainable and healthy development of the rental market .

图片来源

图 2 ~ 7：腾讯街景

图 11：谷德设计网 https://www.gooood.cn/lm-youth-community-china-by-doffice.htm

图 16：腾讯卫星地图

图 13，14，17，19，25 ~ 27：作者自摄

图 2，15，16，28，33：作者自绘

图 8 ~ 11，20 ~ 24，29 ~ 32：拾陌设计工作室

参考文献

[1] 李倩 , 许晓东 . 城中村改造研究热点及趋势 [J]. 城市问题 ,2018(08):22-30.

[2] 徐亦奇 . 以大冲村为例的深圳城中村改造推进策略研究 [D]. 华南理工大学 ,2012.

[3] 程丹丹 . 深圳城中村更新模式对比——以水围村和大冲村为例 [A]. 中国城市规划学会、杭州市人民政府 . 共享与品质——2018 中国城市规划年会论文集（02 城市更新）[C]. 中国城市规划学会、杭州市人民政府 : 中国城市规划学会 ,2018:11.

[4] 佚名 . 深圳最大城中村旧改，15 万深漂人何去何从？ http://baijiahao.baidu.com/s?id=1603201877947873527&wfr=spider&for=pc

[5] 佚名 . 较场尾民宿综合整治——历史与背景 [EB/OL]. 深圳城市设计促进中心 2013-12-01 [2019-02-26] https://www.szdesigncenter.org/lklfLH4carDz3U

[6] 佚名 . 水围柠盟人才公寓，深圳 / DOFFICE—深圳首个城中村人才保障房的诞生 [EB/OL]. 谷德设计网 2017-12-15 [2019-02-26] https://www.gooood.cn/lm-youth-community-china-by-doffice.htm

[7] 马航 . 深圳城中村改造的城市社会学视野分析 [J]. 城市规划 ,2007(01):26-32.

改造后住宅（一）
Remodeled home

改造后住宅（二）
Remodeled home

改造后住宅（三）
Remodeled home

改造后住宅（四）
Remodeled home

1.2

麻磡脚本
——麻磡城中村改造试验及启示

SCRIPT OF MAKAN VILLAGE
EXPERIMENTS AND INSPIRATION OF URBAN VILLAGE RENEWAL IN MAKAN

Tang Bin 唐斌

国家自然科学基金支持项目资助（项目编号：51678126）

我们在后续的教学研究中一直以角色扮演的方式进行着设计讨论，尽可能地以去建筑师化的方式，从政府、村民、使用者和开发者的身份视角判断麻磡的改造；我们所做的设计研究不是以静态的整体规划为出发点，而是可以模糊整体愿景，具体的设计与实施有待于每一步设计研究的深化而循序调整；成果呈现如《罗生门》的视角叠合，从不同视角对麻磡的改造进行解读，在去除先验的价值预设之后，这样的研究更具真实性。

1. 叙事的建立背景

过往对城中村的改造基本上基于一种宏观层面的自上而下，从政府的导向、资金的筹措、项目的运作、规划的流程、成果的预判均显现出终极目标式的图景，成效寥寥。我们的试验探讨从另一个角度做事：既然城中村的土地使用权属于村民或村集体，那么将视角移至产权地块更能平衡土地相关问题；既然城中村目前的现状是对利益最大化追逐的结果，那么何不以各方利益的均衡作为研判的依据；既然城中村人群构成关系复杂，那么社会关系的演变将转化为未来使用者的对话，也将成为贯穿于设计的线索；既然在当下的城中村改造中，政府主导已经成为一种普遍认为的低效手段，那么市场将成为最具优势的判断标尺，它衡量了一切演化进程中的合理与不合理；既然城中村从产生到现在是一个历史的发展过程，那么将其改造定位于阶段性发展更显理性，通过连续性的空间与功能发展逻辑，验证不同时段的演化状态。由此可见，自下而上的城中村改造强调空间单元在市场经济中的自组织行为，同时也受到政策、机会、经济投入等他律的作用，合力导向为一种可能的结果。

另一个采用微观叙事的理由在于我们刻意“去规划”的初衷，并不是常规规划理论与方法在处理城中村改造问题时的失语，而是在惯常的操作模式中，宏观视角虽然能够构建系统性的构架，但错综复杂的宗地关系和个体建筑之间的互动性存在使得这样的整体架构建立有失于效率。相对而言，过程相对于结果更为重要，动态的适应性比静态的终极目的更为可行。

We have been discussing the design in the way of role-playing in the follow-up teaching and research, trying to judge the transformation of Makan village from the perspective of the identity of the government, villagers, users and developers in the way of de-architecting as far as possible. The design research we have done is not overall static regulation. As a starting point, it can blur the overall vision, and the specific design and implementation need to be adjusted step by step with the deepening of study. The result shows the overlap of the perspectives of the movie *Rashomon*, and interprets the transformation of Makan village from different perspectives. After removing the prior value presupposition, such research is more authentic.

1. The dualistic pattern of city and village

In the past, the renewal of villages in cities was basically based on a macro-level from top to bottom. The ultimate goal-oriented prospect was shown from the government's guidance, fund-raising, project operation, planning process and outcome prediction, with little success. Our experiment explores an alternative way from another perspective. Since the land-use right of villages in cities belongs to villagers or village collectives, it is better to balance land-related issues by shifting the perspective to property rights. Since the current situation of villages in cities is the result of pursuing maximum interests, so why not that we rely on the balance of interests of all parties as the basis of research and judge. The evolution of social relations will be transformed into the dialogues among future users and will become a clue throughout the design, since the government-led renewal of villages in cities has become a commonly recognized inefficient means, then the market will become the most advantageous judgment. It measures the rationality and irrationality of all evolutionary processes. Since the villages in city have a historical development process , it is more rational to identify its renewal as staged development, and test the evolutionary state of different periods through the logic of continuous development of space and function. It can be seen that the bottom-up villages in cities' reconstruction emphasizes the self-organizing behavior of spatial units in the market economy, and also receives the role of other disciplines such as policy, opportunity, economic input, and so on.

Another reason for adopting micro-narrative is that our original intention of deliberately "not planning" is not the aphasia of conventional planning theories and methods in dealing with the problem of urban village reconstruction, but inefficiency of the framework from a macro-perspective owing to the complexity in the land parcel ownership and interation among blocks. Relatively speaking, the process is more important than result.It is more practicable to be adaptive with a static ultimate goal.

2. 叙事的设定

1）叙事主体——麻磡村

麻磡位于南山区水源三村中部，紧挨白芒村和大磡村，北临羊台山，南靠西丽水库，处于深圳市的水源保护地范围之内。部分地域性的传统建筑留存至今。在村落的西北部仍然保留致密的传统街巷结构与聚落特征。随着城市的扩展，麻磡成为城市版图包围中的一个异质区域，虽被周遭山体环绕，但村落的产业结构与人口构成已产生质变。传统的农作被第二产业替代，低收入产业工人取代了原住居民，成为村庄的主体。为了增加租金收入，在原居住聚落的西南侧形成了格网状的高密度、高容积的握手楼集群。（图 1）2002 开始的深圳大学城建设将麻磡在规划层面上定位为大学城高校创新圈中的一环。在未来的发展中，逐渐通过退二进三的产业升级，将在校大学生及创新产业初创人员作为主要服务对象。

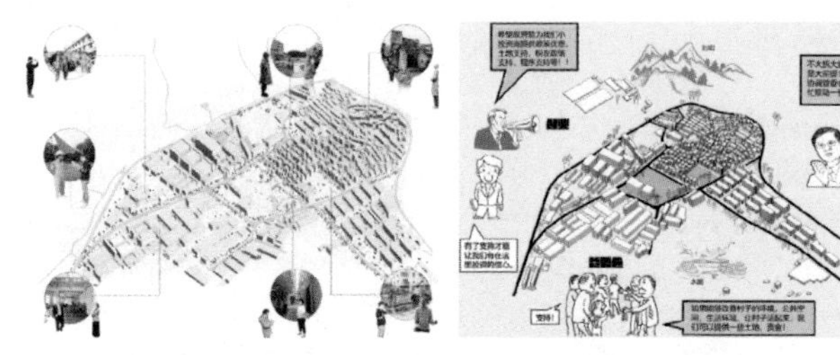

图 1 麻磡村现状空间特征

Fig.1 The spatial characteristics of ma hom village

图 2 三种角色的定位与诉求

Fig.2 The positioning and appeal of the three roles

2）叙事的角色

为了形成对麻磡叙事的多重视角，我们选择四种人群作为叙事的角色：（图 2）

（1）政府

政府关注的是改造的整体成效。在上位政策的引领下制定总体原则，在不破坏整体格局的前提下（不大拆大建），从造纸厂的产业置换和改造开始着手全面改造。

（2）村股份公司

原有集体所有制下所形成的管委会通过成立联营管理公司的形式自发地管理麻磡的更新改造。基于公众利益最大化原则，其定位目标在实现整村产业提升的同时，通过改善整体环境和公共空间品质，实现村子生活的活化。

（3）开发者

开发者是其中的一个潜在力量，他们不从属于麻磡的物质空间但却是麻磡物质空间发展的观察者和行动者，他们代表着市场对于自发性社区（城中村）改造的评价、判断和机会的把握。

（4）创业者

创业者为麻磡未来主要的生活人群，也是未来麻磡经济产出的主体。他们或为周边大学的在读学生，或为刚毕业的年轻人，对于生活工作场所的唯一要求在于较低的支付门槛和与大学园区的便捷交通，并希望得到政府及管委会在土地、空间、税收等方面的支持。

2. Settlement narrating

1) Subject of narrating——Makan village

With the expansion of the city, Makan village has become a heterogeneous area surrounded by the city territory. Although it is enclosed by mountains, the industrial structure and population composition of the villages have changed qualitatively. Traditional farming has been replaced by the secondary industry, and low-income industrial workers have replaced the aboriginal residents and become the main body of the village. In order to increase rental income, a grid-like high-density, high-volume handshakong building cluster was formed in the southwest side of the original settlement. (Fig. 1) The construction of Shenzhen University Town started in 2002 positioned Makan village as a part of the University Innovation Circle in the planning level. The future development , shifting from the labor-intensive industry to service economy, will aim at college students and the enterprise start-ups.

2) Role narrating

In order to form multiple perspectives on Makan's narrative, we choose four groups as narrators: (Fig. 2)

(1) Government

The government is concerned about the overall effectiveness of the renewal. Under the guidance of the superior policy, they should formulate general principles and start renewal from the industrial replacement and transformation of paper mills without destroying the overall pattern.

(2) Village committee

The village management committee formed under the original collective ownership can spontaneously renovate Makan through the establishment of a collective company. Based on the principle of maximizing public interests, its goal is to realize the industrial upgrading of the whole village, at the same time, to revive the village by improving the overall environment and the quality of public space.

(3) Developer

Developers are a potential force in the development. They are the observers and executors of the development of Makan village. They represent the market's evaluation, judgment and grasp of opportunities for the spontaneous transformation of communities (villages in cities).

(4) Entrepreneur

The entrepreneurs are the main body of Makan's future life and economic output. They are either students studying in the surrounding universities or the graduates. The only requirements for living and working places is the lower payment threshold and good accessibility to campus. They also hope to get the support of the government and the management committee in land, space and tax.

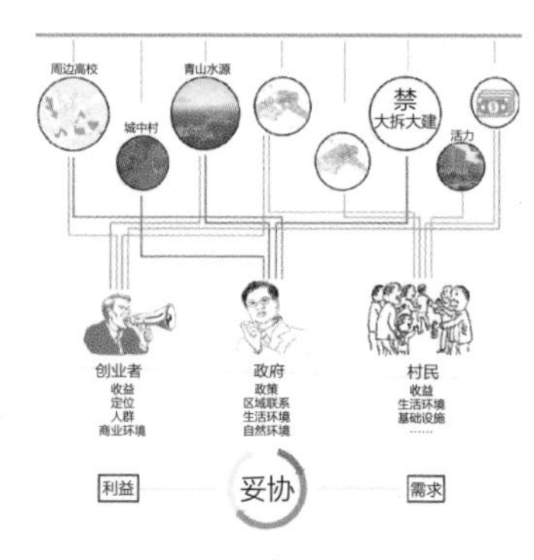

图 3 用合作性利益博弈的方式取得利益的平衡

Fig.3 Use cooperative interest game to get the balance of interests

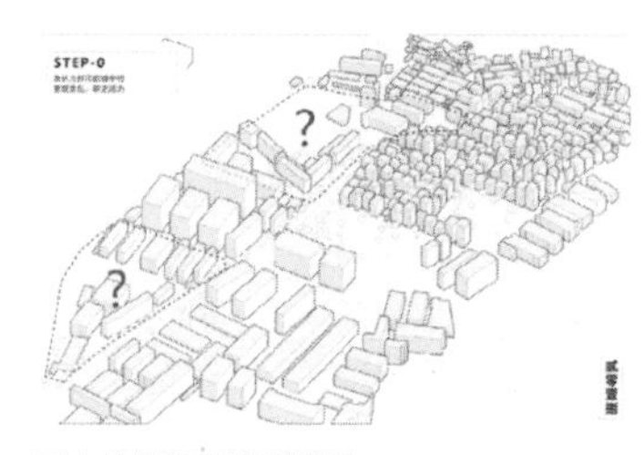

图 4 首位发展地段的选择

Fig.4 First development location selection

3）角色的互动

在我们的叙事里，这四种角色不是相互孤立的存在，他们彼此之间存在着利益的平衡。其中开发者是个外在的力量，我们将其作为一种能动要素；而其余三种角色属于与麻磡整体利益相关的类型，但各自又有着与其自身利益关联的基本要求。如同合作性博弈，我们寻求的不是各方的利益最大化，而是通过相互之间的协作，确立村子改造的阶段和要求。四种人群分别由参与的学生分组扮演，他们在研究的前期建立了交叉讨论的格局，模拟真实状态下的利益博弈，最终达成了一致性的建设性建议和目标（图3）。其中教师的角色被定义为协调者，凭借对各方利益的平衡和调控，协助全组整体推进。

4）叙事的地点设定

麻磡的村庄结构呈现典型的线性特征。外围的二线公路成为过境车流的主要路径，内部的空间单元沿主街依次分布，并按业态成组团布置。村口位置的造纸厂由于具有对环境的污染性，将成为首批整改的工业园区。在传统居住聚落与高密度握手楼区的交界部位是主街的终点，其北侧废弃的学校留下大片的待开发空间，并沿主街保留一些商业服务设施。我们认为这种哑铃式的二分结构，对于村落而言具有一定的生长潜力（图4）。当“哑铃”的两端得以良性发展，中央部分定能成为联动发展的必然选择，并以此为轴，逐步产生向两翼的空间拓展，从而形成空间发展的长程效应。从简单的线性转向更为复杂和开放式的结构类型，从而具备了更强的街区开放可能，具备了自我更新的内在动力，并挖掘了由农业社区向工作 / 生活社区转化的潜力。

5）叙事时间的设定

分时段的设定既为叙事提供了时间线索，也作为分阶段目标达成的依据。为此，我们制定了两种时间的设定标准。第一层次是麻磡整体的发展时间线，以 3 ~ 5 年为一个周期，分为 4 个周期，分别指向启动地块的自我发育，主街沿线的发育，主街两翼的发展，以及村庄的填充过程。我们将研究的重点置于麻磡更新的第一阶段，因为首位发展区域的成败对于整体计划具有重要的导向。在这一阶段进一步将其拆分为三个步骤，以年度为单位进行。通过这样的阶段性更新，一方面微更新行为以持续的小规模投资为前提条件，而持续的条件又以前面的投资获得预期回报为基础；另一方面，小规模的投资和村集体经济体的实力相对应，也适应于初创业者的经济承受能力。在这里我们简化了投资模型，为了排除商业的投机性行为，在外部投资条件良好，并经过良性运作，市场反应积极的情况下，时间周期将得以缩短。

3) The role's interaction

In our narrative, these four roles are not isolated from each other. There is a balance of interests between them. Among them, the developer is an external force, and we regard it as an active factor; while the other three roles are related to the overall interests of Makan village, but each has its own basic requirements from its own interests. Like cooperative game, what we seek is not to maximize the interests of every party, but to sort out the requirements of village reconstruction through mutual cooperation. Four groups of students participated in the group play. In the early stage of the study, they worked in cross-discussion, simulated the real state of interests in the game, and finally reached a consensus of constructive suggestions and goals. (Fig.3) The role of the teacher is defined as a coordinator, assisting the whole group by balancing and regulating the interests of all parties.

4) Narrative Location Setting

Makan village has a linear road network. The outer secondary highway is the main path of transit traffic flow, and the inner buildings are distributed along the main street, and arranged in groups according to their function. Because of its environmental pollution, the paper mill located at the village entrance will become the first industrial park to renovate. The junction of traditional residential settlements and high-density handshake buildings is at the end of the main street. The abandoned schools on the north side of the main street leave a lot of space for development, and some commercial service facilities are reserved along the main street. We believe that this dumbbell-like spatial structure has certain growth potential for the villages. (Fig. 4) With well -developed ends the central part will become the exact choice of the linkage development, and as an axis, gradually expanding to the two wings, thus forming a long-term effect on space development. Makan's spatial narrative will shift from simple linear to more complex and open structural types, possessing a stronger possibility for open blocks and the intrinsic power of self-renewal, and realizing the potential of transformation from agricultural community to work&life community.

5) The Setting of Narrative Time

The division of stages not only provides clues for narration, but also serves as the basis of standards for each stage. The division is under two standards. One is the overall development time line of Makan village, which is divided into four stages with a period of 3-5 years for each. The four stages refer to self-development of the promoter area, development along the main street, the development of the two wings and the development of the whole village. We will focus our research on the first phase of Makan renewal, because the success or failure of the promoter area is the basis of the whole plan. This stage is further divided into three steps, which are carried out annually.

6）叙事内容构成

第一个叙事是村口造纸厂的演变；第二个叙事是村中的社区中心和临近公共地块的变更；第三个叙事是其中的 5 个建筑的自我更新；第四个叙事关于麻磡的整体空间格局转变。每个叙事的方式从基本的问题出发，着眼于现状和可能的资源条件，提案均经 4 个角色的分别诠释，经过利益平衡后，甄选其中综合效益最大的方向作为设计的基础，并在此基础上完成空间的营造。

4 个叙事是麻磡更新的不同片段，也是承前启后的连续过程。故事的讲述（设计的呈现）在脚本的生成中由中观层面开始，再逐步向建筑个体和总体层面延展。一来在价值取向上，我们认为在当前的市场经济下，在完全的自上而下不合时宜的同时，完全听命于市场的导向也会带来公众利益的缺失，合理的整合自上而下与自下而上成为可行之道；二来就操作角度而言，相对于宏观层面过于系统优先和微观层面过于建筑化倾向，选择由地段层面开始更贴近于城市设计操作的有效性。同时，工业厂区组团化的基本格局也决定了中观起步的基本策略，建筑之间的关系从属于最为基本的组团化组织，组团才是与村域系统关联的第一层级。

3. 四个叙事脚本

1）造纸厂叙事

（1）愿景

政府：

原有的造纸厂与水源三村区域作为深圳市水源保护地的定位不符，应及早实现产业转型；作为麻磡改造的“领航计划”，造纸厂的改造应深度结合未来麻磡的总体定位，为未来形成正向引导。

村股份公司：

尽可能地保留原有工业厂房进行改造，节约资金投入，产生更大的效益，吸引外来资金；针对村内商业服务不足的现状，结合沿街界面的改造，增强与村民共享的服务设施。

创业者：

作为初创人员，希望政府和村股份公司能够给予一定的奖励性资助；
提供较多的共享空间和多样性的可变空间，以适应不同的创新产业类型，可以通过自己的参与融入新社区的建设，打造属于自己的创业天地。

开发者：

通过渐进式的改造将其变为城市社区，为年轻人和创业者提供稳定而持续的租售空间；特色化的社区营建，通过特色业态的植入，建立一种生长型社区结构。

Through such periodic renewal, on the one hand, micro-renewal behavior is premised on sustained small-scale investment, while sustained conditions are based on the expected return of previous investment; on the other hand, small-scale investment corresponds to the economic strength of village and initial entrepreneurs. Here we simplify the investment model. In order to exclude speculative behavior of business, the time cycle will be shortened when the external investment conditions are good and the market reaction is positive.

6) Narrative Content Composition

The first narrative is the evolution of Paper Mill at the entrance of Makan village; the second narrative is the change of community centers and the adjacent public plots in the village; the third narrative is the self-renewal of five buildings; and the fourth narrative is about the transformation of the overall spatial pattern of Makan village. Each narrative starts from the basic issues, focusing on the current situation and possible resource conditions, and the proposals are interpreted by four roles separately. After balancing the interests, the direction with the greatest benefits is selected as the basis of the design, and the construction of space is completed on this basis.

The four narratives are different periods of Makan's renewal and a continuous process of connecting the past with the future. The story starts from the medium level, and then gradually extends to the individuals and overall level of whole village. The basic pattern of industrial clustering also determines the basic strategy of the medium-sized start. The relationship between buildings is subordinate to the most basic organization of clusters, which is the first level of association with the village system.

3. Four Narration Scripts

1) Paper mill narrative

(1) Vision

Government:

The original paper mill does not match the requirement to be a water source protection site in Shenzhen, so industrial transformation should be realized as soon as possible. As the "pilot plan" of Makan village renewal, the transformation of paper mills should combined with the overall positioning of Makan village in the future, forming positive guidance through the success of the first step.

Village Committee:

Retain as much as possible the original industrial plant for transformation, save investment, produce greater benefits, and attract foreign funds.

Renew the street interface and increase the shared service facilities to improve the current situation lack of commerce.

（2）利益

政府：

麻磡改造过程中，在厂区的改造中应留有充足的公共设施和场所，提供共享的空间；针对造纸厂的拆迁和改建，提供相关政策方面的支持。

村股份公司：

关注项目实施后在经济利益上的回报，获得比现在出租土地和厂房更高的经济收益；改造中的一些公共设施和基础设施的营造应得到政府部门的支持，通过改善基本条件，使麻磡向成熟化的生活社区转化。

创业者：

只能承担有限的支付成本，但希望得到更多的共享功能，降低空间的使用成本；生活场地和工作场地尽可能接近，周边的生活配套能够齐全，提供城市生活的必备设施；

开发者：

面对麻磡与外界联系不便和相对封闭，以及大学城服务区建设情况不明朗的现状，渐进式的发展策略相对可行，运营的效果决定后续的投入强度；未来的运营应得到来自政府层面的宣传，招商方面应有政府层面的相关政策扶持。

（3）操作

空间分析与策略：

经过空间分析，得出以下的空间调整方法：（图5）

首先通过整理外部空间，形成村头、入口、中心三个不同的开放空间；其次通过拆除较小的厂房和附属设施获得较为连续和完整的外部空间网络，形成内部的步行交通系统；随后在街区外侧边缘通过建设停车楼综合体形成对街区的围合界面；最后改造沿主街和内部街道的建筑内外空间格局，建立二层步道系统，实现造纸厂街区的空间重塑。（图6）

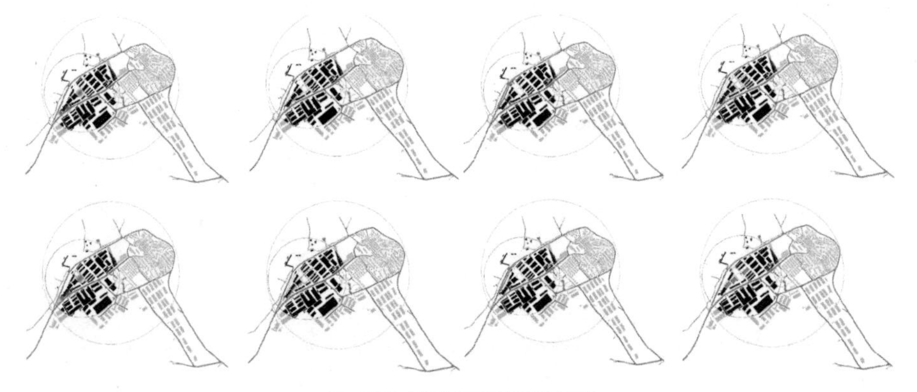

图5 造纸厂区域的问题导向型设计策略

Fig.5 Problem - oriented design strategy for paper mill area

Entrepreneur:

Hope that the government and the village committee can give subsidies.
The transformation should provide more shared space and feasible space to adapt to different types of innovative industries. Through their own participation in the construction of new communities, to create their own business world.

Developer:

Promote stable and sustainable rental space for young people and entrepreneurs through gradual transformation.
Form a growing community structure through characteristic community construction and the implantation of formats.

(2) Interests

Government:

Sufficient public facilities and places should be reserved to provide shared space in Makan village. For the demolition and renovation of paper mills, policy supports should be provided.

Village Committee:

Obtain higher economic benefits than the rental of land and factory buildings. The renewal of public facilities and infrastructure should be supported by the government. By improving the fundamental conditions, Makan village is going to become a mature community.

Entrepreneur:

Require more sharing functions at low cost, easy accessibility between dwelling and working place, and basic amenities.

Developer:

Faced with the inconvenience of contacting with the outside world and the relatively closed situation, a gradual development strategy is relatively feasible, and the effect of operation determines the subsequent investment intensity. Future operation should be publicized, and investment promotion should be supported by relevant policies.

(3) Operation

Space Analysis and Strategy:

Fig. 5 Problem-Oriented Design Strategy for Paper Mill Area.
Firstly, three different open spaces,the front of village, the entrance and the center area, are formed in the external space. Secondly, a more continuous and complete external space network is obtained by demolishing smaller

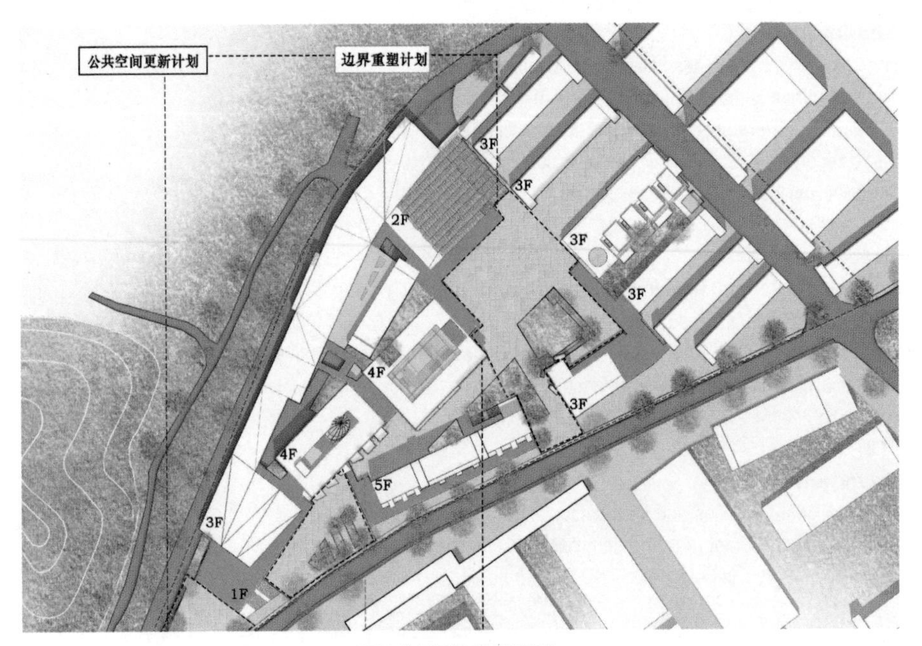

图 6 造纸厂街区的规划意象

Fig.6 Planning image of paper mill block

界面控制：

街区的空间形态控制中，沿主街和外部道路的城市界面具有控制性的作用，形成了街区空间的基本骨架。其中沿外部道路一侧通过新建停车楼综合体方式达成，唯一的底层开口指向车行出入口，一层设置停车库，二层以上布置办公和生活空间。上部的开口具有空间的渗透意义，利于建立空间的暗示并利于内外景观视线的流通。沿主街一侧利用现有建筑进行形体改造，增加底层的灰空间，填补楼栋之间的冗余空间，形成连续性的城市界面。（图 7）随后的界面控制将主要考虑沿内街的空间塑造，重点在于建筑近地面层的空间优化，通过局部架空、抽离、增建等处理，改善建筑之间的空间比例，增强内部街道中的人与人活动的可能，并为楼栋之间交通可达创造条件。（图 8）

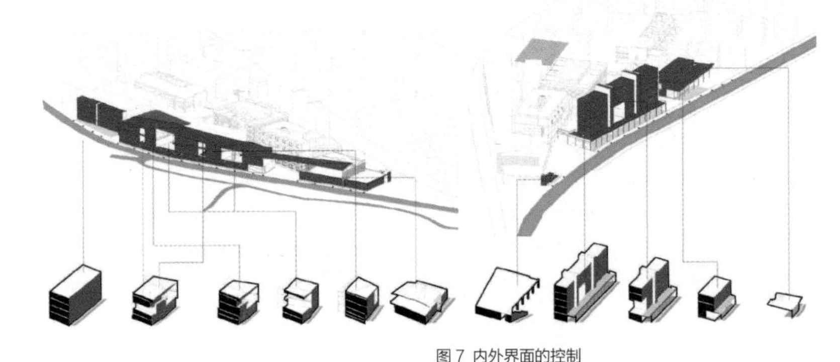

图 7 内外界面的控制

Fig.7 Internal and external interface control

factories and subsidiary facilities to form an internal pedestrian traffic system. Then, an interface, on the outer edge of the block enclosing the block, is formed by a new complex. Finally, the inner and outer space pattern of the buildings will be rebuilt, and the second- storey pavement system will be added to realize the space remodeling of the paper mill block. (Fig. 6)

The control of the interface :

The urban interfaces along the main street and the external road puts up a basic framework for the spatial formation of the block. Along the external road, a new complex will be built. The only opening of the complex on the ground floor faces the entrance of vehicles. The parking garage is set on the first floor while the office and living space are arranged above. The upper opening enbles a better sight. While along the main street, the existing buildings are ultilized, increasing grey zone on the groung floor and filling the redundant space between buildings to form a continuous interface. (Fig.7) The control of interface will focus on shaping the space along the inner street, focusing on the optimization of the ground floor, improving the space ratio between the buildings,enhancing the possibility of human activities in the inner streets, and creating potential for traffic between buildings. (Fig. 8)

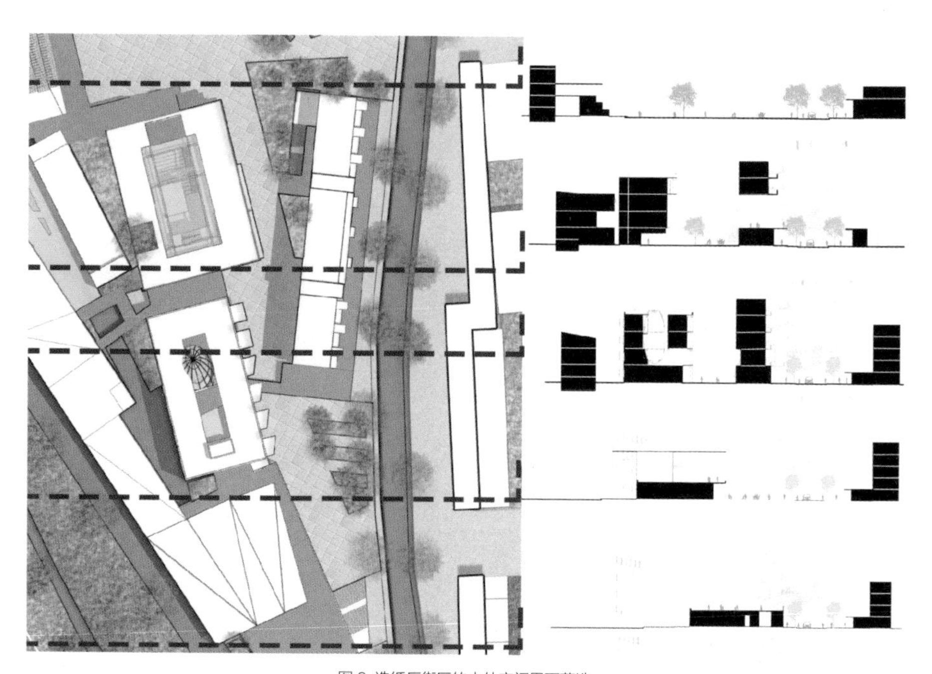

图 8 造纸厂街区的内外空间界面营造

Fig.8 Paper mill block to create internal and external space interface

立体步道与空间组织：

深圳高密度的空间特质决定了较高的空间开发容量，特定的“拥挤文化”标识了高频次的人与人互动，同时也预示了有相对较低入驻门槛的可能。垂直空间发展成为一种必然。不同于一般的城市竖向生长，城中村改造的活力来源带有极强的随机性和偶然性，并随入驻单位的变化而变化，在标准化设计的同时留有一定的“弹性余量”成为一种合理的选择。相对于整齐划一，这种设计中的“非正式”反倒成为一种有趣味的可识别。可借由入驻的创业者参与其中，形成个性化的“片段”。竖向的复合决定了公共交通的竖向叠合、穿插，与楼栋之间的平台、连桥、步廊构件了一个立体的空间意象，更将独立的空间单元整合，提供未来创业者之间交流互动的空间。（图 9、图 10）

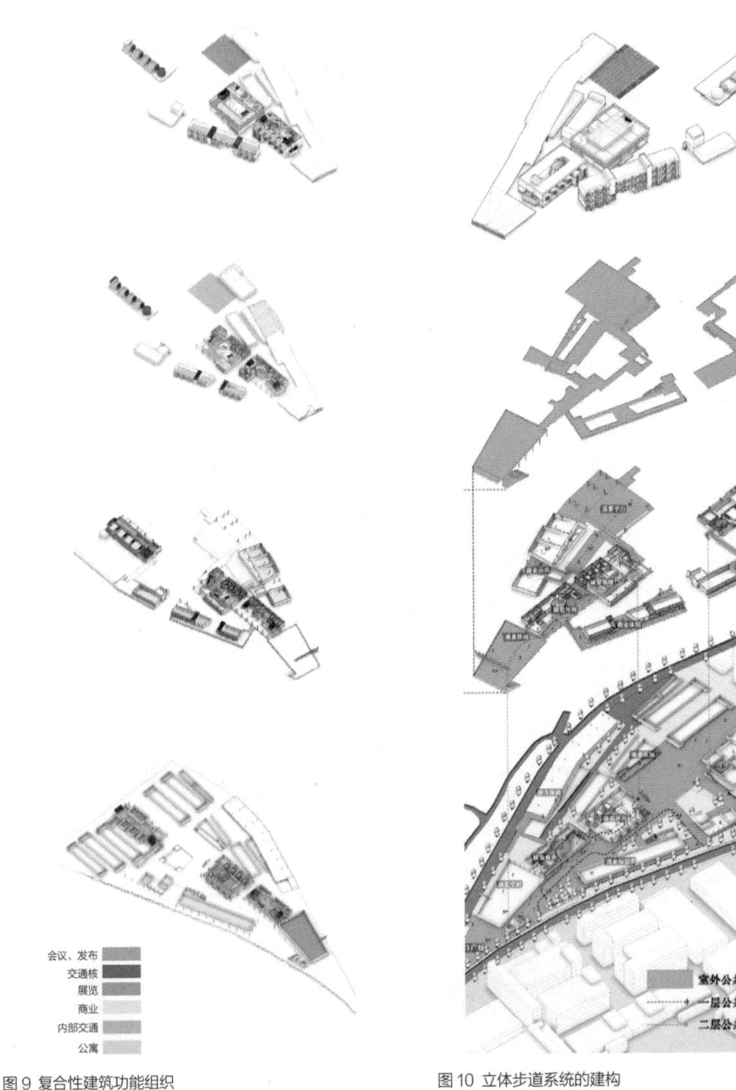

图 9 复合性建筑功能组织

Fig.9 Complex building function organization

图 10 立体步道系统的建构

Fig.10 Construction of stereoscopic trail system

Three-dimensional Walkway and Spatial Organization:

Shenzhen's high-density spatial characteristics determine a high development potential. The congestion of space also indicates the high frequency of human interaction and the possibility of relatively low rental threshold. Therefore, the development of vertical space has become necessary. Different from the general urban vertical growth, the reconstruction of urban village has strong randomness and contingency, and changes with the inserted units. Thus it is better to leave a certain space when standardizing the design. In contrast to uniformity, the informality in this design gives the villages interesting and recognizable characteristic. Through the participation of entrepreneurs, personalized fragments of space can be formed. In verticle direction, platforms, bridges and corridors that intersect among buildings form a three-dimension complex which integrates independent spatial units and provides the entrepreneurs with space for communication and interaction (Fig. 9, Fig.10) .

2) The community center narrative

The renovation of community center is supposed to be carried out after the initial stage of reconstructing the paper mill blocks. With the entrepreneurs moving in, there will be an increase in requirements in the community for commerce, amateur hobbies, professional enhancement and sports. Besides, the supplement of these functions also makes up the deficiencies in the original function system of Makan village to meet the standard of urban community.

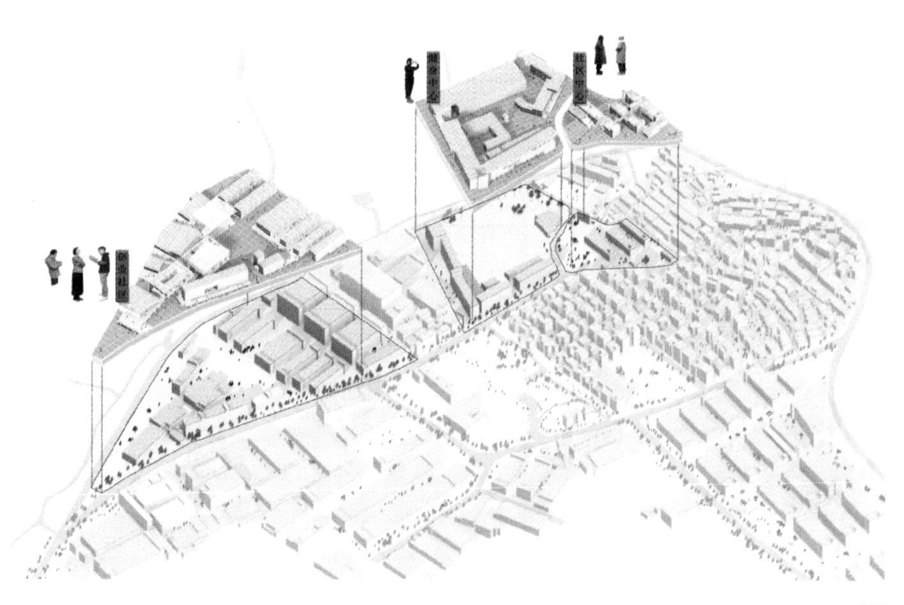

2）**社区中心叙事**

社区中心的改造预设在造纸厂街区获得初步实施成效后启动，随着创业人群迁入，逐步对社区商业配套提出更高要求，同时也对业余的文化生活、专业提升、健身康体等提出需求。这一功能的补充同样填补了麻磡原有功能体系的不足，实现了职能向城市社区的进一步转化。

（1）**愿景**

政府：

借由主街端头地块的改造契机，将其连片打造成社区的商业、文化活动中心；采用以点带面的方式，带动这边废弃地块的建设，进一步完善沿主街的空间界面；

村股份公司：

利用地块的重整，进一步梳理街区的街巷结构和交通环境，坚持街区内部的步行化设计；合理设置街区绿地和开放空间，注重老年人和儿童的户外活动空间营造。

创业者：

随着社群的扩大，对多功能、全时性使用的配套商业、文体娱乐设施的需求进一步增强；共享的展示空间能提供给创业者展示自己的平台，小型的展会能够提供多样化的展示空间。

开发者：

从造纸厂街区的运营中看到双赢可能，希望通过持续的投入得到连续、稳定的经济回报；本着投资可控原则，优先选择改造策略，街区式的空间组织方式将在空间使用层面更为有效。

（2）**利益**

政府：

新中心将为后续奠定坚实的基础，并增强招商引资的吸引力，实现持续性稳定发展的目标；进一步扩展主街的空间整理和界面设计，形成空间发展的两极，进一步增强主街的商业性。

村股份公司：

为入驻者和村民提供更为全面和高水平的综合性设施，配备向现代化综合型社区转化条件；充分利用现有的物质空间资源和闲置土地，在控制投资的前提下实现空间拓展和提升；新街区的改造和开发应充分考虑周边村民的便利性和可达性，尤其关注老年人和儿童人群。

创业者：

在共享经济的主导下，在新的街区中引入新的功能类型和空间类型，增加就业机会；创业者作为独立社会人的权益得到进一步保障，对社区服务设施的建设提出更高要求。

开发者：

从单纯的代建到参与社区的建设与经营过渡，通过持有部分物业获利，增加投资信心；尝试多类型物业的开发，扩大商业规模，期待通过良性的建造、运营获取持续性的回报。

（1） **Vision**

Government:

Taking the chance of the renovationt, the mainstreet will be transformed into a commercial and cultural activity center, stimulating the development of the abandoned area, further completing the interface along the mainstreet.

Village Committee:

The transformation should ultilize the land reorganization, further sort out the street and traffic of the block, and design pedestrian in the block. Also it should reasonably set up green space and open space and pay attention to the environment for outdoor activities of the elderly and children.

Entrepreneur:

With the expansion of the community, the demand for multi-functional and 24-hour commerce and recreation grows increasingly.

The transformation should make room for shared exhibition which can provide entrepreneurs with display platform. Small space feasible to various exhibitions is also required.

Developer:

From the operation of the paper mill block, we notice the possibility of a win-win situation to get stable economic rewards in continuous investment. Based on the principle of controllable investment, the initial task is selecting strategy to re-organize the block more efficient in the usage.

（2） **Interests**

Government:

The new centre lay a concrete basis for future development and enhance the ability to attract investment, maintaining a stable development. The space and interface along mainstreet should be further expanded on both ends of the street, forming a dual development pattern, enhancing the commerce in the mianstreet.

Village Committee:

The transformation should provide more high-qualitied facilities for entrepreneurs and villagers. Also, it should ultilize the original space and lands, and renew the space with controlled investment. The renewal of the block should give full consideration to the accessibility for residents, especially the elderly and children.

Entrepreneur:

New functions should be inserted to increase employment in the block. As an independent social person, entrepreneurs should have further protection in rights and interets, which puts forward higher requirement for the construction of community service facilities.

Developer:

From simple construction agent to participants in the construction and management of the community, developers can increase investment confidence by holding part of the property. They should also attempt to develop multi-type property, expand the scale of business, and expect to obtain sustained returns.

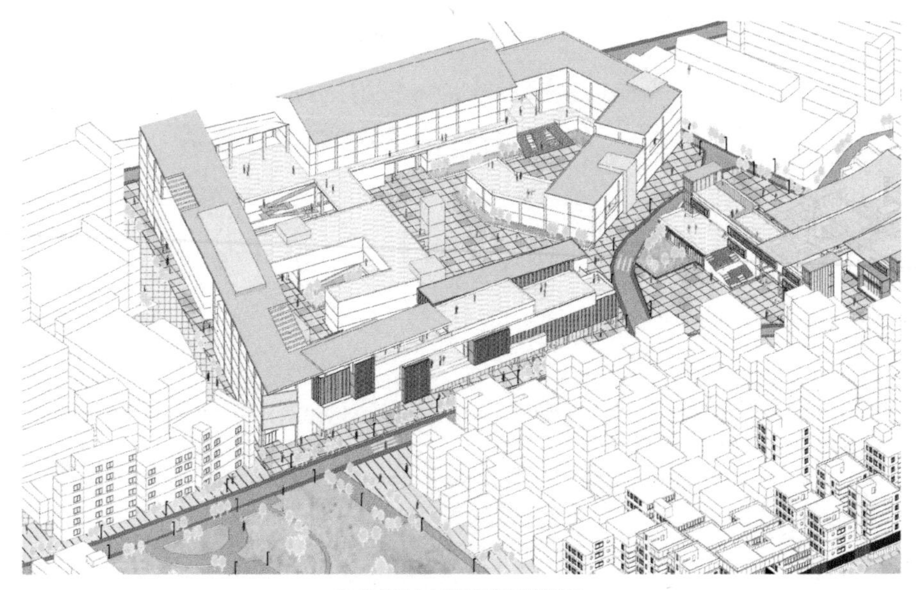

图 11 社区中心地段的空间规划意象
Fig.11 Spatial planning image of community center

（3）操作

空间分析与策略：

该街区由两个相互毗邻的地块组成。其中与村口遥相呼应的一端，具有较强的空间意象性，在实现主街空间主轴的同时还引导南北向的空间序列。另一较大的地块现状仅存沿街设置的建筑界面，空间结构松散无序。设计操作基于场地上述的基本空间问题展开。（图 12）

首先采用围合式方式构建内向型的街区结构，明确区分街道与内部的公共空间，其中外围结构相对体量较为高大，向内部延伸的部分形成多层级的空间围合；其次在两个相邻地块之间梳理外部公共空间的连续性，建立贯穿两个场地，并可向周边延展的步行路径；接着利用内部较为低矮的建筑屋面新建空间联系廊道，形成贯穿子地块的二层步道系统，形成内部空间使用上的连续；最后再通过中心地块屋面的重塑，形成面向主街的空间指向，并成为麻磡最重要的空间公共性标识。（图 11）

界面控制：

社区中心地段的现状界面松散、破碎，尤其是沿北侧过境道路和引道一侧缺少空间限定，通过“织补”的手段实现对地块的填充是必然的设计选择。相对于致密的居住区肌理，工业厂房只留下斑驳的空间组织片段。通过截取典型的建筑尺度作为基本空间参考单元，沿地块周边围合成大小不同的建筑组团；通过连续屋顶等方式，形成意象鲜明的“超级结构体”，以适应于未来 MINI MALL（小型卖场）的引入。面向主街一侧，连续性的空间界面逐渐通过高度的变化，形成指向轴线终端的空间意象。社区中心并行的序列被抽离的大尺度入口门洞打断，暗示着主轴的引入，并借由室外演艺广场的空间折转，转向后侧的居住区道路系统，从而实现了各片区空间关系的关联耦合。（图 15）

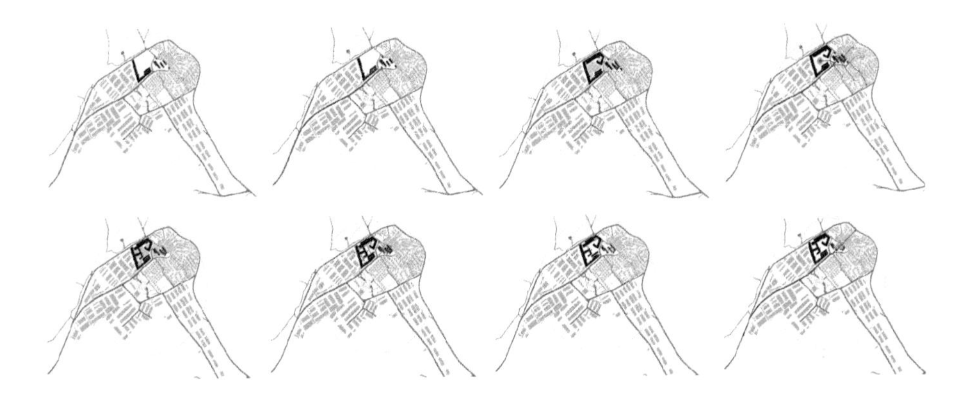

图 12 社区中心地段的问题导向型设计策略
Fig.12 Problem oriented design strategy for community center

（3） Operation

Space Analysis and Renewal Strategy:

The site is divided into two parts by the mainstreet. The part connecting to the enchance has the potential to maintain the mainstreet as the original axis and create a new north-south pathway. In the other part, buildings lies along the steet reasonably organized. Solutions are put forward through the designing process in response to the current problems.(Fig. 12)

Firstly, an introverted space structure is adopted, distingushing street and the inner space. The construction as the boundary is of larger volume and forming a multi-level enclosure. Secondly, out of the introverted structure, a new pathway is designed connecting two parts of the site and extending to the neigbourhood for the continuity of public space. A walkway system will be constructed on top of the low-rise buildings inside the enclosure. Finally, through the remodeling of the roof, the space orientation facing the main street is formed and the inner space becomes Makan's most important public space. (Fig. 11)

The interface is control:

In present situation, spacial organization of the community center area is loose and fragmented, especially along the northern transit road and the side of the approach, lack of space restriction. It is inevitable to stitch it up. Compared to the dense texture of residentrial areas, factories are scattered in the industral area. By intercepting typical building dimensions as the reference of basic spatial units, building groups of different sizes are synthesized along the periphery of the plot, and by means of continuous roofs, a superstructure is formed to adapt to the introduction of MINI MALL in the future. Facing to the side of the main street,along the interface, the height of construction varies, connecting the inner space to the terminal of the axis.The parallel space sequence of community centers is interrupted by a large-scale entrance, introducing the main axis to reach the residential area behind, making and interconnection among the seperate areas(Fig. 15) .

院落空间与功能组织：

界面的生成与各层级内院的生成同步。当厂区向生活社区转变时，各种不同类型的使用必然界定出不同的空间分化可能；同时向内生长的半围合院落进一步修正了周边建筑留下的“超尺度”内部空间，在获得新的使用定义之后，开放性的运动主题广场与各级内院和社区中心地块之间产生了微妙而生动的空间联动。各类商业、文体、休闲、餐饮、培训、创业等城市化的功能模块投入既有和新建建筑的结构网格中，拼贴状的空间意象具象地表明了内部功能的投射。这是属于麻礦的社区“嘉年华”，非正式但充满活力。（图 13、图 14）

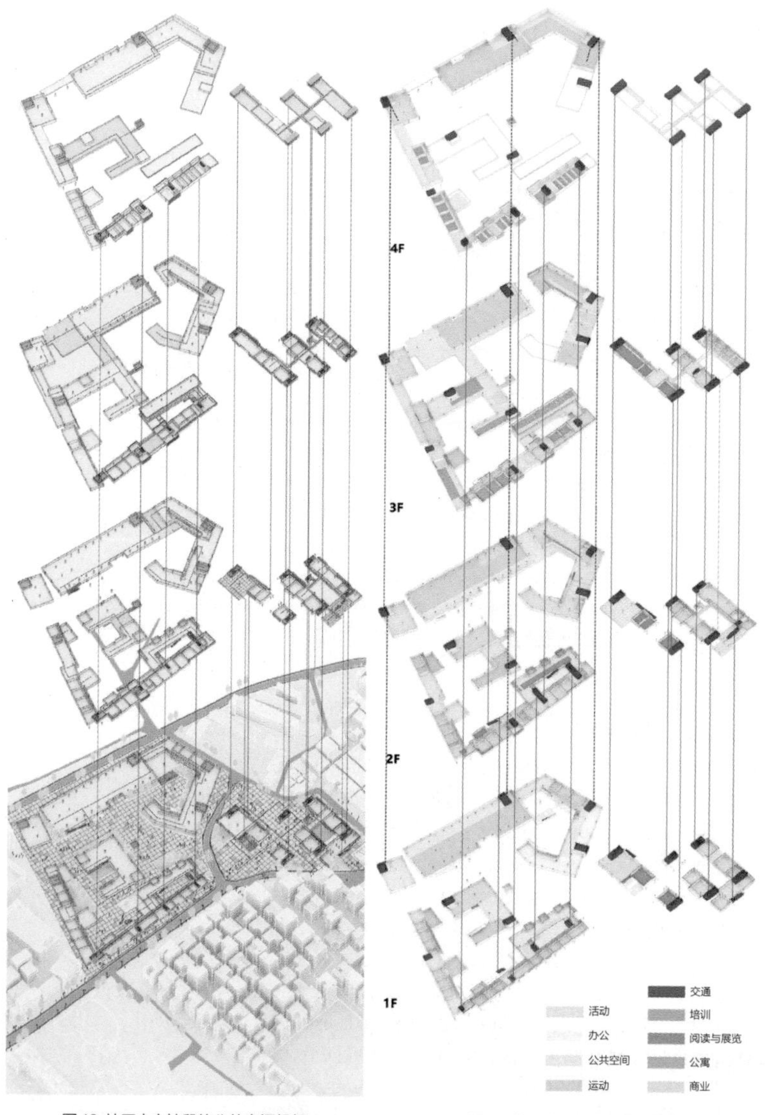

图 13 社区中心地段的公共空间组织
Fig.13 The organization of public Spaces in community centers

图 14 社区中心地段的复合功能解析
Fig.14 Complex function analysis of community center

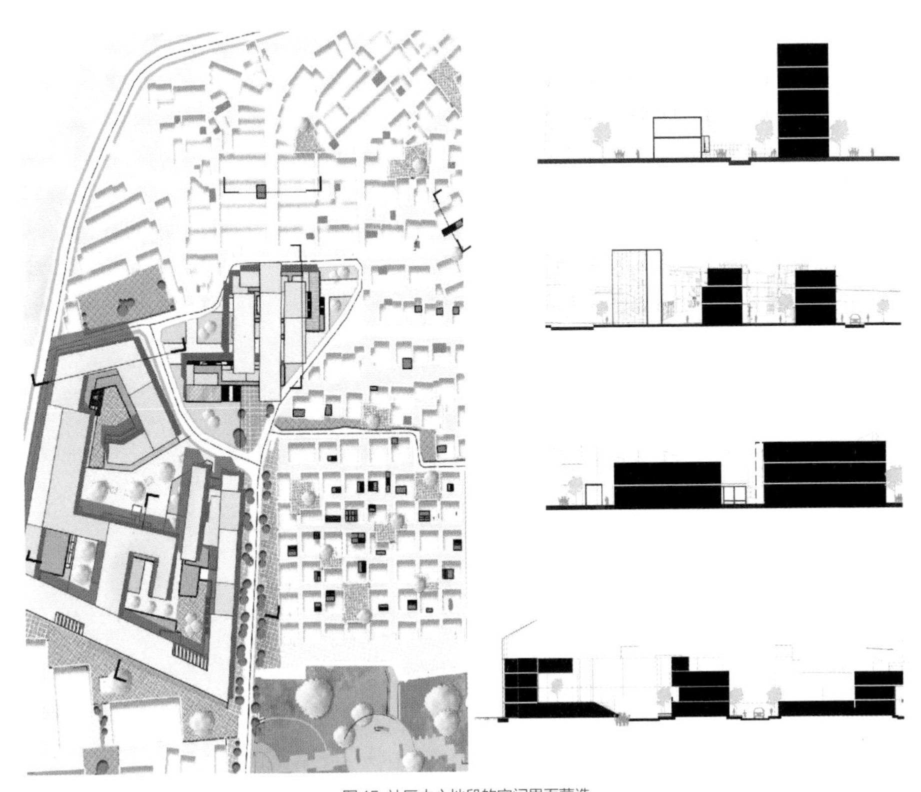

图 15 社区中心地段的空间界面营造
Fig.15 The spatial interface of the community center

Courtyard Space and Functional Organization:

The generation of interface is synchronized with the generation of inner courtyards at different levels. When the factory area changes to the living community, division of space is deternmined by its function. At the same time, it further adjust the super-scale internal to a adorable scale. The transformation of space structure creates an intersection among the sport plaza, courtyards and the community center. Commerce, culture, entertainment, catering, professional institution and start-up companies are inserted in, creatig a vivid, collage-like spacial image. (Fig. 13, Fig.14)

3）5 个建筑的前世今生

建筑的改造设计与中观的城市设计研究同步，在前两个叙事框架建构的同时，伴随着建筑形体与功能的研究，5个建筑的命运悄然改变。

（1）愿景

政府：

尽可能利用现有资源。麻磡现有的工业厂房空间高，规模较大，具有一定的改造灵活性；工业建筑的改造能够改变麻磡村的整体形象，保留历史记忆的同时，也带有鲜明的时代印记。

村股份公司：

解放建筑底层空间。通过在建筑底层通过植入公共空间，能够起到活化内部步行空间的作用；通过建筑的改造与周边的环境设计统筹，彻底改变厂房形象，使其具有视觉吸引力和标识性。

开发者：

充分利用各栋建筑的规模、结构的特点，因地制宜地实现建筑内部空间格局的重塑；私属空间与公共空间的设置应与使用者的生活方式相适应，空间设计与气候适应性对应。

创业者：

提供不同人群之间交流的共享空间，增加空间的流动性和随机组合的可能；共同参与营造自己的“家园”，可参加建筑的微改造行动，可提高社区的生活性氛围。

（2）利益

政府：

政策上鼓励初创者的入驻，并在公共空间的建设方面给予支持，调和村民和创业者的利益；建筑设计应与上位设计保持一致，坚持公共空间部分的设计导则作为建筑设施的控制原则。

村股份公司：

空间设计应能做到大小适中、弹性可调，便于使用、租售，便于回收建造成本；建筑的改造设计应增强使用空间对气候的适应性，降低能源消耗，控制运营成本。

开发者：

尽可能利用现有的建筑结构，按照使用要求进行适应性的改造，降低工程难度和造价；体现空间秩序与既有结构的双重逻辑，突显创造性的空间重塑，提高开发品牌的附加价值。

创业者：

空间设计具有一定的弹性，预留 DIY 的可能性，可通过使用者打造个性化的空间；建筑内部应提供充足的洗衣、简餐、零售、游牧办公等公共设施，增强年轻人之间的互动。

3) The Past and Present of Five Buildings

The renovation design of buildings is synchronized with the study of medium-scaled urban design.

(1) Vision

Government:

Making the best use of existing resources, the space in the existing industrial plant in Makan village is high and has certain flexibility in transformation. The transformation of industrial buildings can put up a new look in Makan Village while preserving the historical memory .

Village Committee:

The ground floor should be activated by inserting public space. The renovation of buildings shall coordinate with the neighborhood and put up an attractive image different from the original factory.

Developer:

Update the inner space uterlizing the original scale and structure. In this process, different lifestyles of residents and the local climate are considered in the first place.

Entrepreneur:

A shared place is required and space with higher mobility and flexibility is prefered. Public participable renewal is welcomed to build up harmony in the neighborhood.

(2) Interests

Government:

Government will conduct policies to welcome more entrepreneus, support the construction of public space and balance the interests among residents and developers. The design of buildings shall follow the principle of public space design.

Village Committee:

Space design should be moderated in size, flexible for usage and rental while easy to recover construction costs. The renovation design of the building should adapt to climate conditions in order to save the costs of operating.

Developer:

The existing building structure should be fully utilized to carry out adaptive upgrade fulfilling requirements of users,moreover, reducing the difficulty and the cost of the project.

Entrepreneur:

The space should be flexibly designed, being able to DIY. The building should should have public facilities such as laundry, snacks, retail, and office for Business Nomads, and so on making space for interaction among young people.

（3）操作

单体 1：

建筑本体为厂区办公建筑，该建筑位于园区内规划步行街道的一侧，原外部空间消极、破碎，且内部空间相对均质，无法产生街区活力。

改造设计首先打开二层空间，与系统化的二层平台连接，然后置入利于自然通风的中庭空间并设置异形的游牧办公体量和主交通结构，再填充单元化的大尺度创客空间以及小尺度艺术家的私密工作空间（图 16）建筑空间的改造与外部公共空间的实施同步，经过整理后的外部街道处于逐步扩大的空域范围内，与园区中心广场保持良好的空间连续。

单体 2：

建筑本体位于造纸厂园区的中心位置，毗邻规划中的中心广场，具有较大的体量和建筑进深。相对于原有厂房封闭、陈旧的建筑形体，消极、无序的外部空间，改造的目标是利用其独具优势的空间位置，创造更好的开放性，并承接二层步道系统的串联。

设计改造采用减法操作，在保留主体结构的基础上，首先面向广场抽离中庭空间，解放步行街一面的封闭墙体；其次在临步行街一侧嵌入阶梯状的活力空间，使得广场空间进一步向建筑内部渗透；最后在背街一侧置入单元化的个人工作室体量，并重新整理广场界面。（图 17）调整后的建筑中庭与二层步道系统无缝对接，并使步道穿行于建筑内外，获得独特的游历感受。步道与建筑中庭均能成为展示、互动的场所，为村民和创业者共享。

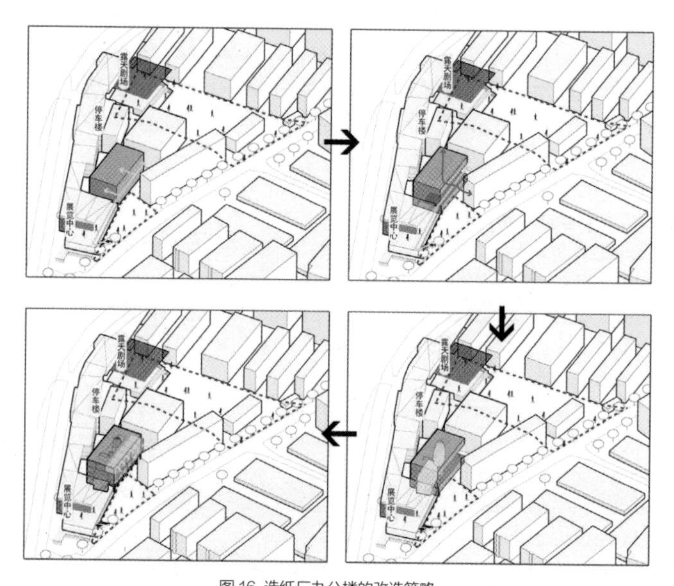

图 16 造纸厂办公楼的改造策略

Fig.16 Paper mill office building transformation strategy

(3) **Operation**

Building NO.1:

The building is originally an office building located beside the pedestrian street of the site. With fragmented exterior space and monotonous interior space structure, it fails to create an active atmosphere.The renovation connects the building with the public walkway system, and introduces in an atrium and normadic offices and corridor system. In the last stage, hackerspace and studios will be inserted. The external renovation synchronizes with the interior. The external space in the shape of trumpet makes a good interaction with the centre plaza(Fig. 16).

Building NO.2:

The building is in the center of the site. Different from the closed and obsolete form of factory buildings, the strategy for building NO.2 is to take advantages of its location, provide better openness, and connect the walkway system to buildings. Subtractive strategy is adopted. The atrium space is extracted, and the enclosed wall on one side of the pedestrian street is demolished. Then, a stair is embedded on the one side of the pedestrian street. Finally, units are placed on the opposite side of the back street. (Fig. 17) The adjusted atrium allows the walkway to weave through the building, creating a unique walking experience. Both the atrium and the walkway can be used for exhibition and communication.

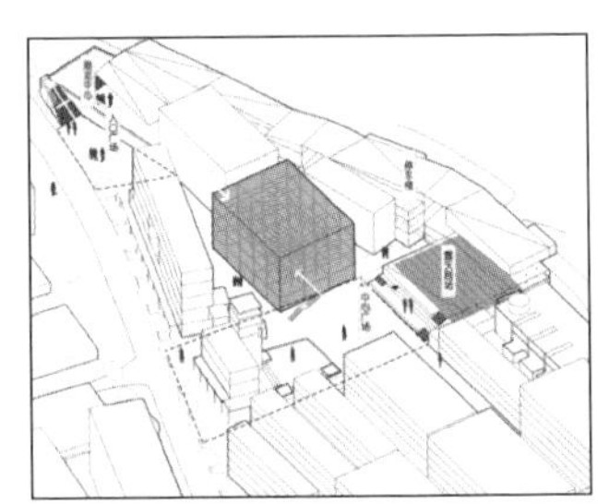
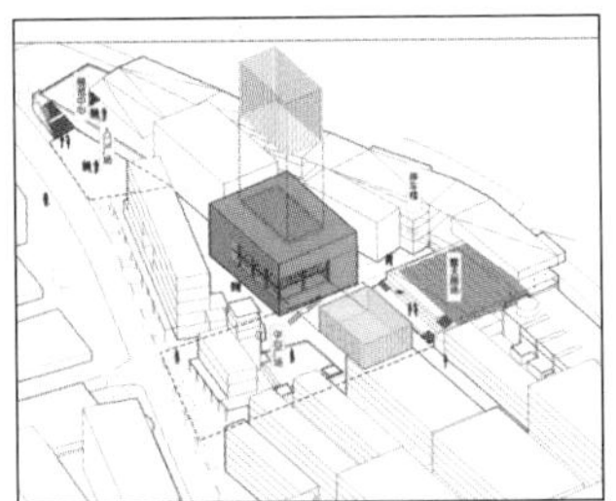
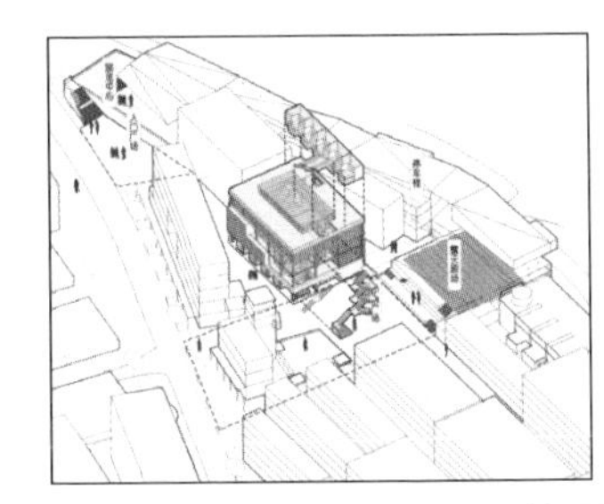
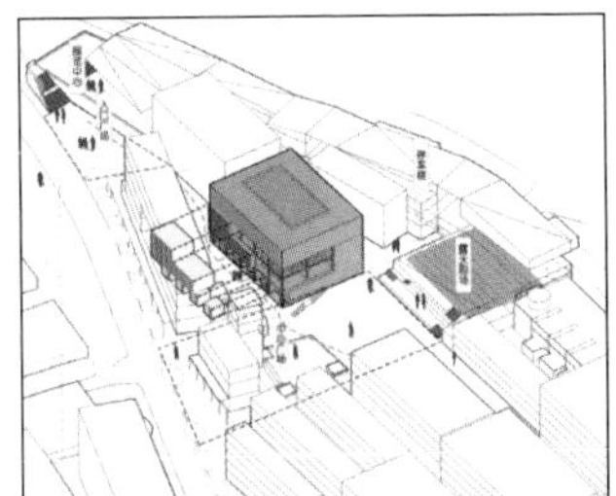

图 17 造纸厂厂房改造策略

Fig.17 Paper mill plant transformation strategy

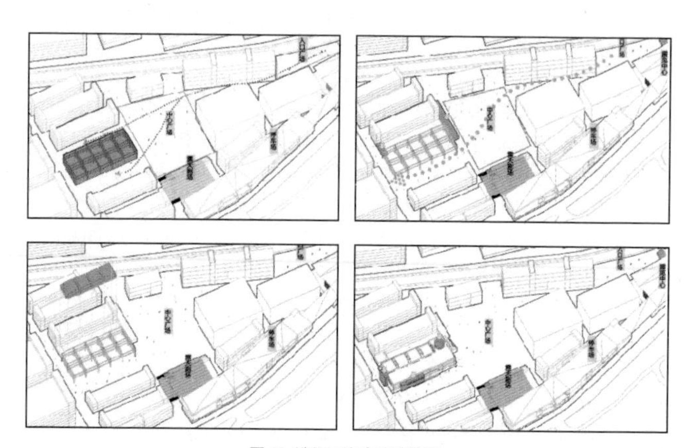

图 18 造纸厂宿舍改造策略

Fig.18 Paper mill dormitory reconstruction strategy

单体 3：

建筑本体原为厂区宿舍，均质排列的建筑山墙面紧邻规划中的广场，在街区研究中，得出以二层步廊串联宿舍组团的空间操作原则，结合步廊一层的公共性设施，可提供未来广场周边的生活活化。改造的目标是将邻近主要步行街道的一栋宿舍改造为兼具展览、会议、文创工作室（含居住）的创意空间。

在设计操作中首先增加山墙面的建筑体量，形成围合式的内院；然后在原有的结构框架基础上置入单元性的功能体，每个功能体块上下错开，分别布置工作室与居住空间，并由独立的垂直交通贯穿；为了增强沿园区主通道一侧的空间公共性，与工作单元并行设置公共空间体量，布置大小展陈空间与会议模块。（图 18）

单体 4：

建筑本体位于主街一侧，当上位设计的格局形成对沿街建筑的改造原则：通过片段的整合促使街廓连续性的生成。改造后的建筑可重新定位为社区的商业服务功能。

经过场地的整理，拆除原场地内一些小型、分散的建筑体量；通过增加连接体量，强化沿主街的界面连续性，并生成内侧的功能性合院；然后将体量研究聚焦于主街界面，在强化水平延展性的同时，突出建筑形体的轮廓变化，形成对社区中心地块的空间指向性，同时在水平体量的组合中有意识地通过材料的定义和体量的脱离，制造“缝隙”，产生“内、外”的空间渗透；随后借助于合院与主体建筑的高度差，产生不同楼层之间的空间联系，并发展为贯穿于整个街区的二层步行系统，制造出内外使用界定明确的立体交通网络；最后采用绿建设计方法，将绿化在屋面和建筑立面上加以投射，创造出绿意盎然的生态社区格局。（图 19）

Building NO.3:

The building was originally the dormitory of the industrial area,with its uniformly-distributed walls adjacent to the plaza. The renewal aims is to renew the dormitory nearby the main pedestrian street into an innovative space for exhibitions, meetings and studios (including residence). It will connect its 2-floor corridor to the public walkway system and stimulate the vitality of the groundfloor by amenities. The volume of the wall will be enlarged to enclose a larger courtyard. Functional blocks will be inserted into the original framework in pairs, overlapping with each other, leaving space for studio and apartments, together forming a unit with independent verticle transportation. Exhibition and meeting place will be installed beside studio. (Fig. 18)

Building NO.4:

The building is located along the main street. The pattern of the upper design forms the principle of the reconstruction of the buildings along the street. The continuity of the street will be enhanced by the integration of fragmented space. The renovated buildings will re-position the commerce of the community.

In site planning, some small and scattered buildings in the original site are demolished. Then, the continuity of interface along the main street is strengthened by increasing the connecting volumes. Functional courtyard inside is generated. Then the study highlights the contour changes of the building and strengthens the horizontal ductility of interface along the street. At the same time, the combination of the horizontal volume consciously creates the "gap" through the definition of materials and the separation of volume, and produces the penetration between internal and extrenal space; then, with the help of height difference between the courtyard and the main building, spatial connection between different floors is well generated. The second-storey pedestrian system runs through the whole block to create a well-defined three-dimensional transportation network for internal and external use. Finally, the green design method is adopted in the roof and facade to create an ecological community. (Fig. 19)

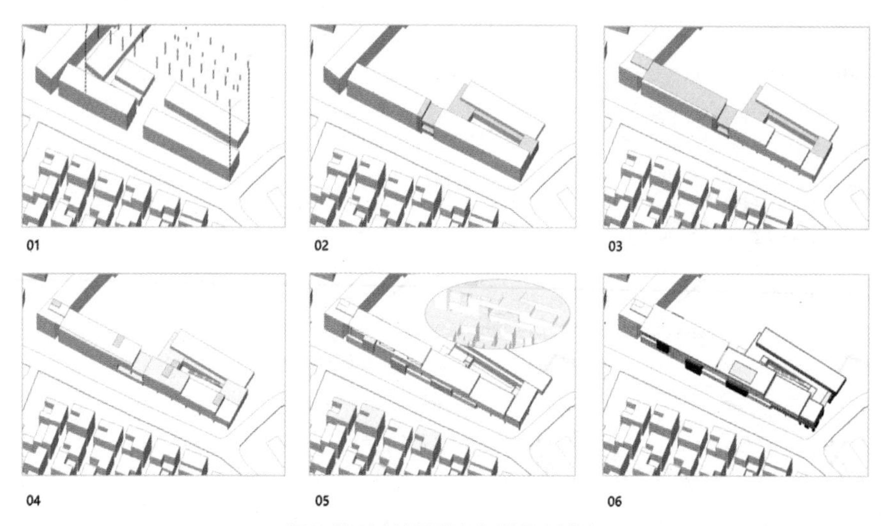

图 19 社区中心地段沿街商业建筑的设计策略
Fig.19 Design strategy of commercial buildings along street in community center

单体 5：

建筑本体原为三栋相互平行且相错的厂区宿舍，因为主立面正对麻磡的空间主轴，因此在总体结构上具有重要的空间意义。实现空间架构上的方向扭转，实现与主轴的对话是改造设计的首要任务。同时由于定位为社区的文化、活动的中心，其内部空间的组织将由均质的单元空间组合转为大小空间的并置。空间的组织方式也将呈现多样的可能，其内外界面也将由于其公共使用目的的出现而得以转变，对外更多地呈现城市街廓的连续性，对内则突出楼栋之间相互嵌套的空间关系，从而通过界面上的加减操作产生空间的耦合。

设计操作首先通过局部的加减和拆除，建立场地的基本空间架构，同时也在场地上划分出户外不同类型和属性的活动空间；其次，界定出场地的主要流线系统，内聚的室外演艺广场面向村子的次要出入口，形成纵横交织的空间结构，贯穿于三栋建筑的立体步道将建筑空间整合；然后借由新建的屋面结构系统完成空间指向性的折转，实现对麻磡空间主轴的呼应；最后通过立面上增减的附属空间体量修正室内使用空间规模，同时也完成了对内外界面的设计。（图 20）

Building NO.5:

The building was originally three parallel and staggered dormitories in the factory area. Because the main plane is directly opposite to the main axis of the space in Makan village, it has important spatial significance in the overall structure. It is the primary task to realize the direction reversal of the spatial structure and the relationship with the axis. At the same time, as the center of community culture and activities, the organization of its internal space will change the original homogeneous combination into juxtaposition of large and small space. Spatial organization will also present a variety of possibilities, and its interface will also be transformed due its public usage, The continuity of urban streets and the spatial relationship between buildings are enhanced, thus creating spatial interconnection through operation on the interface.

The design firstly establishes the basic space structure of the site through partial addition and removal, and at the same time counts out different types and attributes of the outdoor activity space on the site. Secondly, it defines the main flow system of the site, and turn the direction of the outdoor performing plaza to the secondary entrance of the village, forming a interweaving structure within two directions. The complex space structure is then sorted out by a 3 demension pedestrain system. Then, the direction of space is turned by the shape of roof in response to the main axis of the village. Finally, volumes in interface of buildings extend the exterior space and form the elevation pattern as well.(Fig.20)

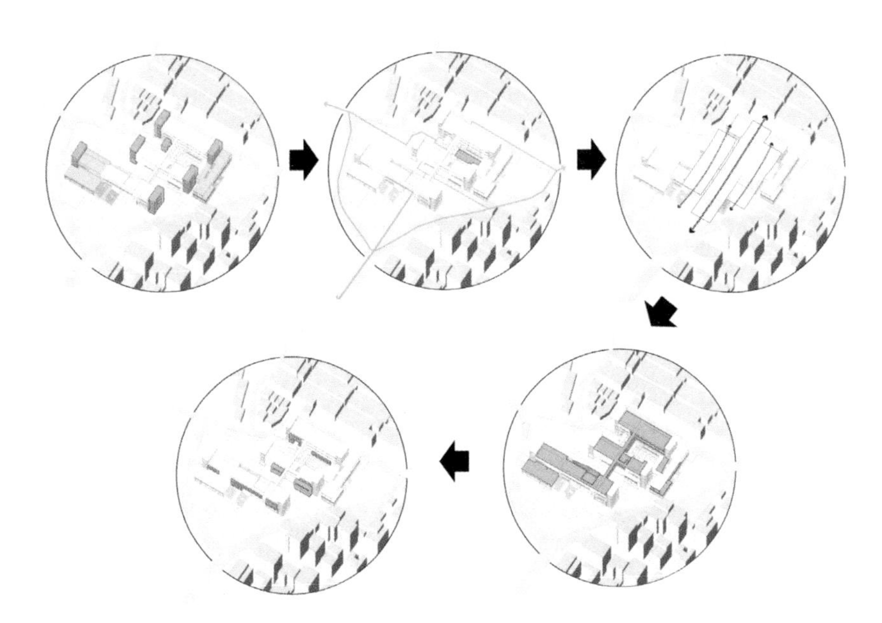

图 20 社区中心建筑改造策略

Fig.20 Community center building renovation strategy

4）麻磡的未来叙事

在完成了三个叙事之后，我们的讨论仍然继续。对于两个发展片区的改造进程过程及发展完成之后将产生的连锁性引发了我们更深层的思考。如果说这种作用是一种良性的机制，那么这种作用必将在麻磡的系统内部产生更为积极的自组织行为，从而导向村落整体的有机更新。因为各种利益的博弈作为内在的发展机制，也因为哑铃结构的建构触发了麻磡演化的引线，这种演化将按照潜在的规则持续，从而实现关于麻磡的未来叙事。

（1）愿景

政府：

通过渐进方式对现有工业厂区进一步改造和调整，实现第二产业向第三产业的转型；依托于周边良好的山地景观条件，发展衍生产业，使麻磡成为新型的综合性城市社区。

村股份公司：

借由这样的更新程序，重新调整村落的空间结构，使其具备城市社区的空间框架和功能配置。

开发者：

未来麻磡在主导业态基础上，存在着多元发展的可能，多元化的功能结构与投资体系可行；寻求麻磡的特色化经营之道，通过向综合性社区的转变，创造更大的经济价值和社会价值。

创业者：

随着麻磡逐步融入城市体系，将出现更多的创业类型，空间的有序化将成为重要的空间特征；各项设施的运作应得到进一步的优化和提升，使创业者真正享有“城市”生活。

（2）利益

政府：

打造新兴的高智产业和服务聚落，通过社区联动，全面提升麻磡的空间品质与建成环境；逐步完善基础设施建设，保证麻磡物质空间到社会环境的全面提升。

村股份公司：

全面保障村民和创业者的基本利益，优化投资环境，提供良好的社区配套；各发展片区之间应保持协调和一定的相关性，确保整体利益与局部利益的兼顾。

开发者：

鼓励建立共管共赢的模式，参与各方均能从综合效益的最大化中获取利益；投资与建设的主体由村集体主导向政府、村集体、开发商联营的方式拓展，突显成效。

创业者：

针对创业机构的规模和发展状况提供更多样的选择；完善商务功能，提供更为优越的创业空间；片区式的发展是相对更为有序的发展策略，各片区之间应建立必要的联系，便于使用。

4) The future Narration of Makan village

The transformation and development of the two zones trigger our thoughts. If this function is a beginning mechanism, it will produce more positive self-organizing behavior in Makan village, which will lead to the organic renewal of the village as a whole. This evolution, promoted by different interests and a dumbbell shape growing pattern, will continue with the rules of former narrative to make up the future of Makan village.

(1) Vision

Government:

Through gradual transformation and adjustment of existing industrial factories, transformation from secondary industry to tertiary industry can be realized.

Village Committee:

On the basis of the construction of pilot sites, the optimal spatial development route is formulated to realize the village renewal plan of Makan village.

Developer:

The development of Makan village has a good investment prospects. As a new high intellectual concentration area in the future, it has a certain long-term investment value.

Entrepreneur:

With the gradual integration of Makan village into the urban system, more types of entrepreneurship will emerge, and the organization of space will become an important spatial feature. Functions will be improved to meet with urban life standard .

(2) Interests

Government:

High-intelligence industries and service settlements should be ceated, through community interaction, comprehensively improve the spatial quality and construction environment of Makan.

Village Committee:

Coordination and relevance should be maintained among the development zones to ensure that both overall and local interests are taken into account.

Developer:

The continuous development of Makan village should be based on clear overall objectives and continuous policies. Encouraging the establishment of a win-win model of co-management, all participants can benefit from the maximization of comprehensive benefits.

Entrepreneur:

Diversified choices for entrepreneurial institutions basing on scale and development conditions should be provided. Business functions should be optimized to provide more superior business space. Regional development can be conducted in an organized way but still requires interation among seperated regions for convenient usage.

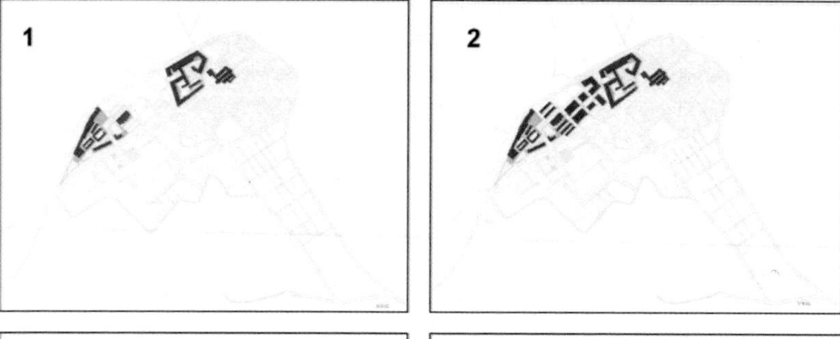

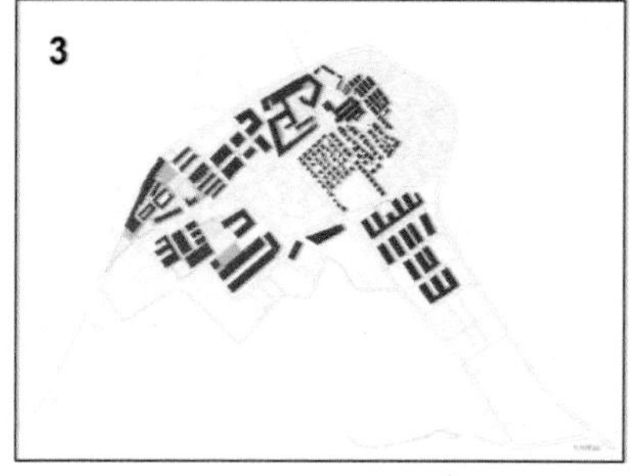

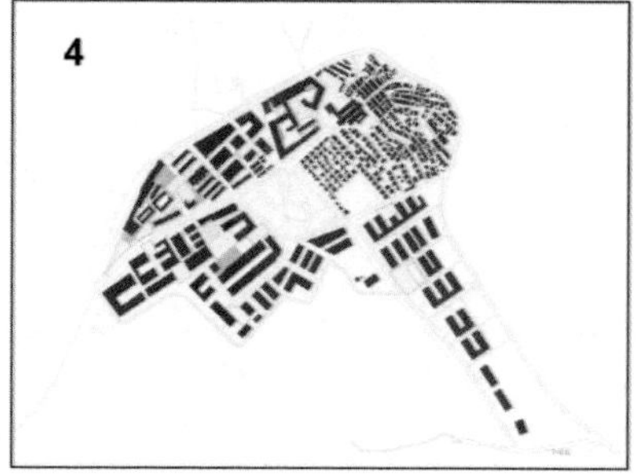

图 21 麻磡空间的渐进式发展

Fig.21 The development of makan space is gradual

（3）操作

分步走的策略为麻磡的前期建设带来了良好的开端，后续的村落改造虽可吸引更多的市场关注和经济投入，但市场导向下的有序建设仍将作为麻磡叙事的主旋律。

空间发展：

“哑铃”状的分片区发展暗示了中段的发展潜力。一方面，因两个组团的磁力作用，在自组织模式下，相关的资源条件优先向该片区周边吸引，逐步形成了片区边缘的拓展；另一方面，两个片区对主街界面的严控，以及麻磡商业职能向主街一侧的集聚使得沿主街一侧的空间改造将获得更多的系统涨落，从而具备更多的发展机会。当这一阶段初步达成，主街以北片区将形成贯穿于各地块之间的另一条隐性发展线索，它串联了造纸厂地块和社区中心两个地块的主要公共空间网络，从而形成了与主街并行的另一系统分支。

主街在麻磡村落空间结构中的主导性决定了其向周边辐射的能力。空间发展的第二阶段将沿着主街的另一侧展开。在不远的将来，在主街的另一侧，可发展出 2 个组团，它们因为邻近已发展地段和麻磡的中心公园，将成为第二优选发展区域。此时，潜在的公共空间线索进一步跨街延伸，最终导向中心公园这一村域核心开放空间。

最终，空间的填充作用持续发展，潜在的空间线索最终形成贯穿于整个村落的另一条重要的空间序列，串联起村域中所有的功能组团，新的空间结构也由此产生。（图 21）

(3) Operation

The step-by-step strategy has brought a good start to the early construction of Makan village. Although the follow-up village reconstruction can attract more market attention and investment, the orderly construction under the market orientation will remain the theme of Makan's narrative.

Space Developing:

The development of dumbbell-shaped patches implies the development potential of the middle section. On the one hand, due to the magnetic effect of the two groups, under the self-organizing mode, the related resources preferentially attract to the periphery of the area, and gradually form the periphery. On the other hand, the strict control of the interface of the main street in the two areas and the concentration of the commercial functions of Makan village promote the space transformation along the side of the main street. There will be more fluctuations in the system and more opportunities for development. When this stage is preliminarily reached, the area north of the main street will form another hidden development clue which runs through the blocks. It links the blocks the paper mill block and the community center, thus forming another branch system parallel to the main street.

The dominance of the main street in the spatial structure of Makan village determines its ability to radiate to the surrounding area. The second stage of space development will be carried out along the opposite side of the main street. In the near future, two new groups can be developed on the other side of the main street, which will become the second preferred development area because of their proximity to the developed areas and the Central Park in Makan village. At this time, the potential clues of public space further stretch across the street, and ultimately lead to the central park, the core open space of the village.

Ultimately, the filling function of space continues to develop, and the potential spatial clues eventually form another important spatial sequence throughout the whole village, which links up the functional groups in the village area, and the new spatial structure also emerges. (Fig. 21)

Transition of the Structure Type:

The spatial structure of the village corresponds to the development of the village. The present spatial structure of Makan village corresponds to its relatively simple spatial texture distribution. The spatial form of Makan centres on the central park, and the village, except for the northeastern residential area, enclosed by various industrial factories. A "L" shape road links all parts of the village. Commercial and public facilities gather in the village at the northern end of the road. Therefore, the linear feature is the initial structural prototype of Makan village.

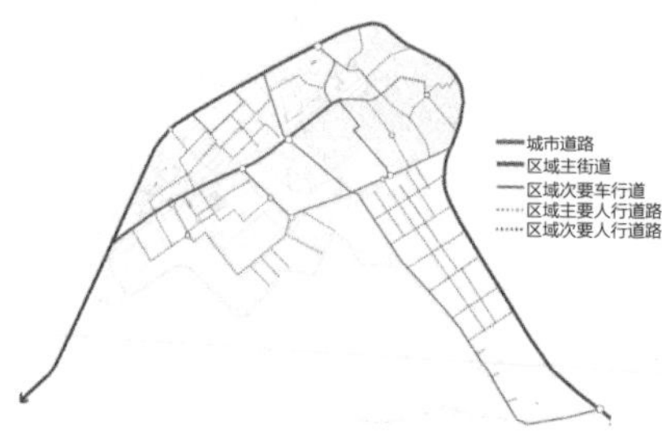

图 22 麻磡的路网与道路结构演化

Fig.22 Road network and road structure evolution in ma hom

结构转型：

村域空间结构与村域的发展状态相对应。麻磡现在的空间结构与其相对简单的空间肌理分布相适配。麻磡的空间形态以中心公园为核心，除东北部为居住组团外，其余均被各种工业厂区包围。一条“L”形的道路将村域各部分串联，其中北端聚集了村里的商业和公共设施，成为主街。因此，线性特征是麻磡的初始结构原型。

结构的生长伴随着空间层级的丰富而改变。主街的“一”字形结构随着社区中心次要出入口的建立演变成了“十”字形结构；贯穿于各功能片区的连续空间路径演化为一个闭合的环形，并连接了中心公园与各组团的核心公共空间。该环交接于社区中心地块，强化出主轴的空间意象。社区中心以东的传统居住街区道路路网密集，经过梳理，形成放射状的聚合，同样交集于社区中心。如此，麻磡的空间结构演化为两轴一环一中心的结构特征。在此主结构之下，还存在着各组团空间的次级结构，空间的层次性得到进一步加强，一个适应于未来社区化的空间框架由此产生。

交通与路网格局：

过境的二线公路提供了基本的对外交通条件，为了阻止外来车辆的进入，村口设置了控制闸机，并在南部环线设置村里的公共停车场。所有的村内机动交通依赖于“L”形的基本道路骨架。大量的工业厂区在其内部形成自身的内部路网，仅在主街留有进出通道。与之形成鲜明对照的是居住组团致密的步行道路呈蛛网状蔓延，缺乏有效组织。这种简单层级的路网模式显然不能适应于麻磡未来城市社区转向下的交通模式。

当代城市社区的交通组织基本原则是根据交通流量和属性的区别将道路交通分级设置。其中过境交通经由二线公路疏解；通过各片区的停车配建设施，将汇入村内的机动交通在外围按片区消解，减轻村内的机动交通负荷；主街通过道路断面设计控制机动车速度，提倡慢行化设计。穿行于各功能片区的公共空间环道实施全步行设计，创造人性化的步行空间。现代社区路网格局的另一个特征是密路网小街区。随着工业厂区的逐步解体，原有的厂区内部道路具备了公共化的可能，原本以厂区为单位的交通组织方式转变为以楼栋为单位，格网状的步行道路提供了便捷的可达性，同时也为社区的公共化提供了条件。经过梳理后的传统住区也通过路网的分级设计，整理出放射状的道路主干，融入规则的道路格网。（图 22）

The growth of structure changes with the enrichment of spatial hierarchy. With the establishment of the secondary entrance and exit of the community center, the "I" shape structure of the main street has evolved into the "cross" shape structure, and the continuous path running through the functional areas has evolved into a closed loop, connecting the central park and the core public space of each group. The ring is connected to the community center plot, which strengthens the main axis. The road network of the traditional residential blocks is in high density. After sorting out, it forms a radial aggregation, which also intersects in the community center. In this way, the spatial structure of Makan evolves into the structural characteristics of two axes, one ring and one center. Under this framework, there are secondary structures of each group space, and the spatial hierarchy has been further strengthened. Thus, a spatial framework adapted to the future community emerges.

Traffic and Street Pattern:

The secondary highway provides basic external traffic conditions. In order to prevent the entry of exotic vehicles, control gates are set up at the village entrance and public parking lots also are set up on the Southern Ring road. All vehicles depends on the "L" basic road network. A large number of industrial factories form their own internal road network in their own interior, leaving access only with the main street. In contrast, the dense pedestrian roads of residential groups spread cobweb-like and lack effective organization. This simple hierarchical road network model obviously can not adapt to the future traffic mode in the transformation of Makan village.

The basic principle of traffic organization in contemporary urban communities is to classify road traffic according to the difference of traffic flow and attributes. The transit traffic is cleared by the secondary highway while the motor traffic flowing into the village is cleared by the parking allocation facilities in each area, so as to reduce the motor traffic load in village. In main street, speed of motor vehicles is controlled through the design of road sections, and slow-moving design is advocated. The public circular road passing through each functional area is designed to create humanized walking space. Another feature of the modern community street pattern is the dense network of small blocks. With the gradual disintegration of the industrial site, the original road inside the factory area has the potential for publicity. Traffic organization mode originally based on the factory area has changed in accordance with the building. The grid-shaped pedestrian road provides convenient accessibility, and also provides conditions for the community publicity. After sorting out the traditional residential areas, through the graded design of road network, the radial road trunk is sorted out and integrated into the regular grid. (Fig. 22)

4. 麻磡叙事后序

麻磡叙事是我们对中观视角楔入城市街区改造的研究试验。城中村带有很强烈的，由乡村脱胎而来的印记，但在城市系统的浸染下依旧带有从发生到发展的过程性。在市场经济下，在大拆大建的模式被摒弃后，何种方式才是处理城中村（更甚或推演至一般的城市传统街区）更新的有效方法，是值得仔细斟酌的。麻磡试验无疑是在一种自我设定的语境中进行的尝试，当将城市的演化归为一种内外规则合力作用下的必然结果，这样的一种规则建构以及演化的模拟，无疑就具备了某种必然性。

1）一场自下而上的更新试验

对于城中村，抑或是城市老旧区域的更新，传统粗放式操作带来城市“焕然一新”的同时，引发了诸多物质层面、文化层面、社会层面的问题。从长远的视角上看，传统的改造更新策略，实则在以公众利益作为代价换取“看上去很美”的表象，表面的浮华光鲜之下的暗潮涌动，其实是对“城市使生活更美好”这一基本准则的拷问。相对于我们已经习以为常的“甲方思维”，从需求出发，从利益的各方出发，能真正地贴合市场经济下城市更新的终极目标，能在可持续、渐进式的途径下提供一条新的思路。正如任何一个物质空间的产生和发展都是在各种内在机制和外部约束下的自然演化，其更新的历程与未来状态的呈现也不应由个人意志决定，回归于市场，用“看不见的手”及多重利益的均衡作为评判的手段，似乎有着不同于以往的操作优势，尤其是面对极为复杂的城市物质空间系统与社会关系时。

4.Post-narrative Preface of Makan Experiment

Makan's narrative is our research experiment on the transformation of urban blocks from a medium-sized perspective. The village in the city bears a strong mark, which comes from the birth of the village, but it still has a process of occurrence and development under the influence of the urban system. In the market economy, after the mode of large-scale demolition and construction is abandoned, what is the most effective way to deal with the renewal of villages in city worths careful consideration (or even in the general urban traditional blocks). The Makan experiment is undoubtedly an attempt. When the evolution of a city is attributed to an result under the combined action of internal and external rules, such an organized construction and evolution simulation are undoubtedly inevitable.

1) Experiment on Bottom-Up Update

For the renewal of villages in cities, or old urban areas, the traditional extensive operation has brought about a "refreshing" of the city, at the same time, it has caused a lot of material, cultural and social problems. From a long-term perspective, the traditional strategy of renewal is actually to exchange the appearance of "looking beautiful" for the public interests. Compared with the the thoughts from Party A perceptive, which we have been accustomed to, starting from the needs and interests of all parties, it can truly meet the ultimate goal of urban renewal under the market economy and arouse new thoughts on a sustainable and progressive way. Just as the emergence and development of any material space is a natural evolution under various internal mechanisms and external constraints, the course of its renewal and the presentation of its future state should not be determined by individual will. It seems to be advantageous returning to the market, using the "invisible hand" and the balance of multiple interests as the means of judging, especially when it comes to extremely complex urban material space system and social relations.

City is a complex giant system. Villages in cities or urban blocks are only one segment of it, and they are still attached to this complex relationship network. For complex systems, the traditional mode can only solve problems unilaterally, but can not reveal the internal mechanism and the relationship among subsystems, thus unable to provide a complete solution, wasting efforts. Under the theory of complex systems, Edgar Morin's new paradigm of science gives new perspectives and methods of thinking. If the vision of villagers, users and developers give us a more dimensional perspective, if their pursuit of their own interests in the market economy becomes the reality that we must face, then the design research has indeed needs reflection. Without interference, the interests of these groups are achieved through cooperative game between them. But unlike the complete market orientation, the state-owned policy of our land determines that the government will always be the maker of the core rules, and its fair interests will always be superior to other related relations. In this way, we can not simply deduce urban problems

城市是复杂的巨系统，城市村或城市街区仅是其中的一个片段，但依旧依附于这种复杂关系网络。对于复杂系统而言，传统的操作方式只能单方面解决问题而无法揭示内在的运作机制和子系统相互之间关系，无法完整地提供解决策略，从而多数的努力归于无效。在复杂系统论下，埃德加·莫兰（Edgar Morin）的科学新范式给予了新的思考视角和操作方法。如果来自村民集体、使用者和开发者的愿景将我们从单一的设计研究拓展至更多维的角度，如果他们在市场经济下对各自利益的追逐成为我们必须面对的事实，那么设计研究的确到了需要反思的地步。在不加干扰的状态下，这些群体的利益通过彼此之间的合作性博弈达成。但不同于完全的市场导向，我们土地国有的政策决定了政府永远是核心规则的制定者，其代表的公允利益永远凌驾于其他相关关系之上。这样，我们就不能简单地从埃德加·莫兰的范式进行城市问题的推演，而是在自组织和他组织之间寻求一种可能的契合点。

城市系统的自组织依赖于系统内部各要素之间的能动和联动。按照元胞自动机理论，系统任何一个阶段的状态是由系统内各元素状态叠合而成，并随着组成要素的变化而改变，规则成为有限预测和判定系统状态的关键参量。因而，完全人为的规划或个人意愿的呈现只能是一厢情愿的梦呓。合作性博弈的前提是将利益多方的地位和需求对等，这也是市场公允条件下的一个必要假设和社会致力达成的目标。（虽然在现实中话语权的掌控永远存在着强弱的区别）在某种程度上，最终的成果与其说是设计的结果，不如说是按照这样的利益互博而得出的某种相互妥协。

在有序发展的系统中不存在相对的静态。对于城中村抑或城市街区而言，活化的途径就是通过人为制造的涨落，激化系统的规则演变。在涨落存在的情况下，相对的平衡被打破，在规则的驱动下，依靠系统元素之间的相互作用达成新的动态均衡。当这种激化作用为正向时，系统得到进化，当量变累积到结构性的改变，则系统完成质变的跃迁。在麻磡试验中，对造纸厂地块和社区中心的改变，相对于麻磡村域的更新来说就是一种通过事件的引入，产生的连锁反应。随着资源在局部区位的集中，形成明显的涨落关系，从而产生了对主街沿线的直接带动，并进而促使全村域的结构改变。在此过程中，涨落发生点的选择至关重要。首选系统中的关键节点相较于其他区位有着更为深层的作用能效，这就如同针灸相对于全身系统的作用机制。其次，对于作用区位的选择应能以产生远程效应为优选，并能通过本区域的成果激发涨落作用的传递，以达成远端的作用。在麻磡试验中，一个基本假设就是各方对利益的追逐是一个常态，并随着改造规模的扩大产生更大层面的预期，使得市场作用下的正向引导产生连续效应。

from Edgar Moran's paradigm, but seek a possible balance.

The self-organization of urban system depends on the dynamic and linkage among the elements within the system. According to the theory of cellular automata, the state of any stage of the system is composed of the states of each element in the system, and changes with the changes of the components. Rules become the key parameters to predict and determine the state of the system. Therefore, the complete artificial planning or the presentation of personal wishes can only be wishful dreams. The premise of cooperative game is to balance the position and demand of multi-stakeholders, which is also a necessary assumption under the condition of fair market and the goal that society strives to achieve. To some extent, the final result is not the result of design, but rather a compromise based on the mutual benefits.

There is no relative static state in the system of orderly development. For villages in cities or urban blocks, the way of activation is to intensify the rule evolution of the system through artificial fluctuations. In the presence of fluctuations, the relative equilibrium is broken, and driven by rules, a new dynamic equilibrium is achieved by the interaction of system elements. When this intensification is positive, the system evolves, and when the equivalent changes accumulate and stimulate structural changes, the system completes the qualitative transition. In Makan experiment, the change of paper mill plots and community centers is a chain reaction through the introduction of events compared with the renewal of Makan village. With the concentration of resources in local areas, there is a clear fluctuation relationship, which directly drives the main street, and then promotes the structural change of the whole village. In this process, the choice of fluctuation occurrence point is very important. Compared with other locations, the key nodes in the preferred system have a deeper effect on energy efficiency, which is similar to the mechanism of acupuncture relative to the whole body system. Secondly, the selection of action location should be able to produce long-range effect as the best choice, and can stimulate the transmission of fluctuation effect through the results of the region, so as to achieve the far-end effect. In the Makan experiment, a basic hypothesis is that the pursuit of the interests of all parties is a normal state, and with the expansion of the scale of renewal, there will be a greater level of expectations, which will make the positive guidance under the action of the market produce a continuous effect.

Completely bottom-up urban renewal is difficult to implement under China's current land regulations. Even in the society of land privatization, it will encounter inefficiency and can not be completely controlled. In an open system, the evolution of the system is determined by the relationship between the elements. It is impossible to delude into the exact state of a certain moment, but can only make a limited prediction. On the other hand, the planned urban development in China is implemented in every link of urban construction management mechanism, which determines that the construction of urban areas can not exist independently beyond control. From this point of view, we can't hope that the conditions and process of the generation of

完全自下而上的城市更新在中国当前的土地制度下很难推行，即便在土地私有化的社会里，也会遇到效率低下，无法完全受控的状态。在开放的系统中，系统的演化由元素之间的相互关系决定，无法妄言某一时刻的确切状态，只能做有限度的预测。另一方面，我国的城市发展计划性和规划工作贯彻于城市建设的每个环节，这样的管理机制决定了城市片区，甚至于每一个楼栋的建设均无法超越管控而独立存在。从这个角度而言，我们不能寄希望于过去那种完全生长性的村落或城市生成条件与历程，能够做到的是在自上而下与自下而上之间的某个中间态。这种由机制引发的设计操作已经与传统的规划或城市设计手段划清了界限，对麻磡而言，以微观思考、中观操作、宏观导出作为思考的基点是正确的选择。

2）从旧金山海滨试验到麻磡试验

从麻磡试验的四个脚本不难看出亚历山大（Christopher Alexander）旧金山海滨试验的影子。他将该试验记录在《城市设计新理论》中，在他看来，建筑的生成过程与城市空间形态的形成息息相关，二者之间的互动对于城市空间整体性的形成具有重要影响，而整体性是决定城市空间品质最为重要的环节。促使建筑生成的核心规则是在渐进式的发展过程中保持更上一级层次城市空间的整体性。

在这个试验中，亚历山大采用了自下而上的方式实现了城市空间形态的生成。他采用的方式同样是指导学生扮演每一个楼栋的策划与设计的工作。城市就在一栋栋建筑的逐步填充中完成了空间的增殖。无疑，这是一种更为精确和细微的城市设计工作。但不难看出它与麻磡试验的区别所在。

首先旧金山试验是针对城市空间发展与空间形态生成进行的研究，其视角是建筑师的，而非规划师的。其探讨的内容和成果均从，也仅从建筑形态的生成逻辑与未来产生的空间潜力展开，因此是单视角的，也是形态决定论的。麻磡试验是立足于利益的相关性进行讨论，更关注于空间发展的内在动力和机制，关注于业态类型对于社区转变的作用及潜在发展远景，甚至于建筑师在工作的大部分时间，其专业身份是退场的。因此麻磡试验是多视角的，非建筑决定论，也非规划决定论。

其次旧金山试验是从建筑到群体的有机发展过程。为了找到创建日益增长的城市整体性发展所需要的各种法规，亚历山大提出了一套初步法则，它们体现了实际发展的过程，与城市日益发展要求相吻合。这是在一种先验的基础上进行的一个有目的的行为，试验的目的是验证法则的有效性，设计在其中的作用至关重要。麻磡试验同样是建立在规则的基础上，但这个规则并未以书面的形式规定，而是隐性的潜藏于每个参与者的价值判断中，与建筑或城市片段的空间形态虽无直接关联，但影响并导致空间形态的生成。设计工作具有更广义的范畴，甚至可以说最重要的工作并不是由设计所决定。

villages or cities in the past can be achieved . A middle stage is most possibly achievable. This kind of design triggered by mechanism has made a clear distinction with traditional planning or urban design methods. For Makan village, micro-thinking, meso-operation and macro-derivation are the correct choices for thinking.

2) From San Francisco Coastal Experiment to Makan Experiment

It is not difficult to see the elements of Christopher Alexander's San Francisco coastal experiment from the four scripts of the Makan experiment. Alexander recorded the experiment in "New Theory of Urban Design". In his opinion, the process of building formation is closely related to the formation of urban spatial form. The interaction between them has an important impact on the formation of urban spatial integrity, which is the most important link to determine the quality of urban space. The core rule of building generation is to maintain the integrity of urban space at a higher level.

In this experiment, Alexander adopted a bottom-up approach to achieve the formation of urban spatial form. His approach is also to guide students to play the role of planning and design for each building. The city has completed the proliferation of space in the gradual filling of buildings. Undoubtedly, this is a more precise and subtle urban design. But it is not difficult to see the difference between it and the Makan experiment.

Firstly, the San Francisco experiment is a study on the development of urban space and the formation of spatial form. Its perspective is of an architect, not a planner. The contents and achievements of the discussion are from the perspective of the logic of the formation of architectural form and the potential of space in the future. Therefore, it is single perspective and form determinismatic. Makan experiment is based on the relevance of interests to discuss, pay more attention to the intrinsic motivation and mechanism of space development and the role of the type of industry in community renewal and potential development prospects. Even in most of the time when architects work, their professional identity is withdrawn. Therefore, Makan experiment is multi-perspective, not architectural determinism, nor planning determinism.

Secondly, the San Francisco experiment is an organic development process from architecture to community. In order to find the laws and regulations to create a growing city, Alexander put forward a set of preliminary rules, which embodied the actual process of development and accorded with the requirements of urban development. This is a purposeful behavior based on a priori ground. The purpose of the experiment is to verify the validity of the rule. The design plays an important role in it. Makan experiment is also based on rules, but this rule is not stipulated in written form, but hidden in the value

再者在旧金山试验中是一个从微观出发的设计操作，概由土地产权问题引起。在土地私属的前提下，每个独立地块分属于不同的业主，全权拥有对土地上部物业的主导权利。因此，总体设计就是在空间层面上处理各独立地块的土地使用。唐泽恩学派就（Conzen school）认为被历史学家和建筑工作者称为城市肌理的概念，实际上是由城市平面、土地分割的地块模式和在地块分属的三维物质结构所组成。这里的肌理是城市中由各种建筑的形体特征在城市空间形态上的综合反映。而麻磡试验是从中观出发，因为村中的土地以集体所有为主，这些土地上由村集体建造的工业产区由多栋建筑组合而成，空间关系的讨论主要集中在小区域内部，而与村域空间无关。因而对建筑单体的讨论实际上并不能触动村子的整体系统。

最后在旧金山试验中，建筑师是在场的。每个参与者的角色就是一个独立的建筑师个体，在完成自己地块项目的同时，还需兼顾于整体片区的内在空间逻辑生产，以及周边项目的影响。而从麻磡的四个叙事中，我们仿佛看到建筑师或规划师的缺位，每个叙事中的设计操作仅是对各方愿景和利益平衡后的因果。从自组织方式上看，在系统的有序演化中，规则是一切动作产生的根源。一旦系统的涨落形成就在公约的条件下完成而无须更多的干预，一旦干预则成了对系统规则的改变。因此我们可以认为，“设计”首先是一种程序（program）的设计，它建立了一种共同工作和讨论的机制，从而使得工作得以展开；其实“设计”是一种协调关系的制定，它建立了一套评价的体系和判断的标准，从而使得彼此的工作既做到顾全大局又着眼特定权益。只有设计回归自身的语境，才成为一种操作的方法，才体现出其本体的意义。

3）建筑师的作用讨论

建筑师作用的讨论一直贯彻于试验的过程中。一方面我们质疑自己的学科背景是否能有效地完成这样的角色转换，另一方面也在考虑在体制内的建筑师制度是否适应于这样的操作方式。或许这是一项更加艰巨的工作，或许这些工作根本不在本专业的考虑范畴，无须杞人忧天。但随着工作的推进，我们越加坚信，这样的工作唯有具备强烈责任感和使命感的建筑师才可胜任，并不是我们对规划专业的鄙视链使然，而是建筑师才能真正在地性地讨论微观的基本问题。同时，我们又时时提醒自己将自己从建筑师的身份中抽离，因为当摆脱了职业的习惯性思维，我们才能真实地以另外的视角和价值观去看世界。而我们要做的，正是在这些相互交织的线索和视角中找寻相互叠合的部分，那里隐藏着麻磡真正的未来。

judgment of each participant. It is not directly related to the spatial form of buildings or urban fragments, but affects and leads to the formation of spatial form. Design has a broader scope, and even the most important work is not decided by design.

Furthermore, the San Francisco experiment is a micro-based design operation, which is caused by land property rights issues. Under the premise of private ownership of land, each independent block belongs to different owners and has the dominant right to the superstructure of land. Therefore, the overall design is to deal with land use at the spatial level at the independent block level. According to Conzen School, the concept of urban texture, is actually formed by urban plan, land division pattern and three-dimensional material structure on land subdivision. The texture here is a reflection of the urban spatial form by the physical characteristics of various buildings in the city. The Makan experiment starts from the middle view, because the land in the village is mainly owned by the collective, and the industrial production areas constructed by the collective are composed of many buildings. The discussion of spatial relationship is mainly concentrated in the small area, but has nothing to do with the village space. Therefore, the discussion of building units can not actually touch the whole system of the village.

Finally, in the San Francisco experiment, the architects was present as an individual. While completing his own plot project, he takes the internal space logic production of the whole plot and the influence of the surrounding projects into account. From Makan's four narratives, we can see the absence of architects or planners. The design operation in each narrative is only the balance of visions and interests of all parties. From the point of view of self-organization, rules are the root in the orderly evolution of the system. Once the fluctuation of the system is formed, it will be completed under the conditions of the convention without more intervention, and once intervention, it will become a change of the rules of the system. So we can think that what to design is firstly a program, which establishes a mechanism for joint work and discussion. In fact, design is the formulation of a coordinated relationship, which establishes a set of evaluation system and criteria for judgment, so that each other's work can be done. We should not only take the overall situation into consideration, but also focus on specific rights and interests. Only when design returns to its own context can it be carried out in practice.

3) Discussion of Architect's Role

The discussion of the role of architects has been carried out throughout the experiment. On the one hand, we question whether our academic background can effectively accomplish such a role transformation. On the other hand, we are also considering whether the architect system is suitable for such a mode of operation. Perhaps this is a more arduous job. Perhaps these jobs

建筑师在面对这样的更新项目时，因为需要处理较多的建筑个体，以及建筑背后的利益问题，往往讨论最多的集中在形态层面，其余的不能进入他们习惯的“舒适区”。为此，北尾靖雅（Yasunori Kitao）曾提出过一种基于总建筑师（MA）与地块建筑师（BA）的互动协调方法来处理城市设计和群体建筑设计中的相关问题。在其理论著作《城市协作设计方法》（Collective urban design: shaping the city as a collaborative process）一书中，提出了协作城市设计的概念。北尾靖雅将其定义为一种城市规划的方法，其目的是应用这种方法创造和谐、个性化和环境友好型的群体建筑形态。MA—BA模式建立起一种局部与整体之间的广泛联系，使局部设计受到整体的控制，同时又能对城市设计、城市规划既定的整体建设目标产生部分的修正，是一种注重实效和过程的设计运作方式。广泛的协商贯穿于设计运作的全过程，有效、及时的信息反馈和设计信息的共享，为整个设计团队的协作创立基础。虽然这种方法致力于解决建筑设计中的形态相关的处理方法，但其内在机制值得我们参考。

在麻磡试验中，MA 的角色是相对模糊的，抑或可以说这是一个相互商定以后的虚位，不是一个具体的参与人或指导教师，而是一种价值观念的存在。BA 既是每个参与人，也可以理解为一种抽象，因为他们不仅代表着建筑师的个体，也代表着他们利益代理人身份。这样的相互讨论，就远远超出了北尾靖雅设定的建筑形态层面，而深入到业态选择，相关权利与义务，建造成本的分担，运营的效果与评价，外部空间基本设计准则等更为广泛的层面。当这些规则被明确的制定，后续的设计工作则成为一种参与人之间的自觉。因此我们认为北尾靖雅提出的 MA-BA 模式在我们的工作中依然有效，只是转换了话语的语境。

对于建筑师作用的第二重讨论其实是一个“老生常谈”之举，即建筑师在城中村或城市街区改造中的地位、话语权及起到的作用。在我国已有的城中村或城市街区改造中的模式有三种：政府主导、开发商主导和村集体主导。在第一种模式中，建筑师的作用其实是政府意图的执行者，主要完成前期策划、协助制定实施计划和规划设计工作。这时建筑师的工作只是集中在一个相对集中阶段，脱离最终的服务对象，话语权受到极大的制约。在第二种模式中，开发商处于关系链条的核心，政府和村集体在某种程度上成为开发商可供利用的资源和讨价还价的对象，建筑师成为开发商的某种代言人，将其利益的追求付诸设计，始终脱离不了背后的利益团体操控，话语权自然也得不到自由的彰显。在村集体主导模式中，村集体需要面对筹措资金、制定计划、落实项目、评估效益等多方面的责任，作为一个基层经济组织，村集体自然不能完成这样的专业化程度极高的工作。聘请执行力强、专业化程度高的建筑师或团队作为坚实的技术后盾是一种可行之道。这时建筑师就成为一种代理人身份，他们依靠自身的技术能力，处理从策划、规划、设计、实施中的相关问题，直面终端使用者，并与各个相关部门保持协作，因而具备了最大的话语可能，也具备了最积极的主动性。

在麻磡试验中，建筑师身份的模糊在于其不再局限于村集体主导下的角色扮演，而是在多重身份的重叠。一方面作为村集体的代理人，

are not in the scope of professional considerations at all. There is no need to worry about them. However, with the progress of our work, we are more and more convinced that only architects with strong sense of responsibility and mission can be competent for such work. It is not because of our disdain for planning profession, but because architects can truly discuss micro-basic issues in the local context. At the same time, we often remind ourselves to separate ourselves from the identity of architect, because when we get rid of the habitual thinking of profession, we can truly see the world from other perspectives and values. What we need to do is to find the overlapping part in these intertwined clues and perspectives, which is hidden, and which is the real future of Makan village.

Facing of such renewal projects, architects often focus on the form level. But in fact, because they need to deal with more individual buildings and the interests behind the buildings, all of these beyond form can not enter their habitual "comfort zone". To this end, Yasunori Kitao has proposed an interactive and coordinated approach based on the master architect (MA) and the block architect (BA) to deal with the problems related to urban design and group architecture design. In his theoretical book "Collective urban design: shaping the city as a collaborative process", the concept of collaborative urban design is proposed. Yasunori Kitao defines it as a method of urban planning. Its purpose is to use this method to create harmonious, personalized and environmentally friendly group building forms. MA-BA model establishes a wide connection between the parts and the whole, which makes the local design controlled by the whole, and at the same time can partly modify the overall construction objectives of urban design and urban planning. It is a kind of design operation mode that pays attention to actual effect and process. Extensive consultation runs through the whole process of design operation. Effective and timely information feedback and sharing of design information create the basis for the collaboration of the whole design team. Although this method is devoted to solving the form-related treatment methods in architectural design, its internal mechanism is worth our reference.

In the Makan experiment, the role of MA is relatively vague, or it can be said that it is a vacancy after mutual agreement, not a specific participant or instructor, but the existence of a value concept. BA is not only a participant, but also an abstract one, because they represent not only the individual architect, but also the agent of their interests. Such mutual discussion goes far beyond the architectural form level set by Yasunori Kitao, and goes deep into the broader aspects of mode selection, related rights and obligations, cost sharing, operation effect and evaluation, basic design criteria of external space, etc. When these rules are clearly formulated, the subsequent design

深圳城中村概览

城中村缺乏活力

城中村建筑风貌差

城中村建筑密度高

城中村居住状况差

work becomes a kind of consciousness among participants. Therefore, we believe that the MA-BA model proposed by Yasunori Kitao is still valid in our work, but it only changes the context of discourse.

The second discussion on the role of architects is actually "cliche", that is, the status, discourse power and role of architects in the renewal of villages in cities or urban blocks. There are three modes in the transformation of villages in cities or urban blocks in China: government-led, developer-led and village collective-led. In the first mode, the architect's role is actually the executor of the government's intention, mainly completing the preliminary planning, assisting in the formulation of implementation plans and planning design work. At this time, the architect's work is only concentrated in a relatively concentrated stage, leaving the ultimate service object, the right to speak has been greatly restricted. In the second mode, developers are at the core of the relationship chain. To some extent, the government and village collectives become the resources available to developers and bargaining objects. Architects become some spokespersons of developers. Their pursuit of interests can never be separated from the control of interest groups behind them. Naturally, the right of speech can not be displayed freely. In the village collective dominant mode, the village collective needs to face the responsibility of raising funds, making plans, implementing projects, evaluating benefits and so on. As a grass-roots economic organization, the village collective naturally can not complete such highly professional work. It is feasible to employ architects or teams with strong execution and high professions as solid technical backup. At this time, architects become an agent. They rely on their own technical ability to deal with the related issues from planning, design to implementation, facing the end users directly, collaboration with the relevant departments and keeping, so they have the greatest possible discourse and the most positive initiative.

In the Makan experiment, the ambiguity of architect's identity is no longer confined to the role-playing under the leadership of village collectives, but the overlap of multiple identities. On the one hand, as the representative of the village collective, the architect fulfills the relevant functions entrusted by his duties. On the other hand, he also has three perspectives of government, developers and end-users to judge the relevant affairs from their respective angles and coordinate the conflicts of interests between them actively. Therefore, it is better to define it as a coordinator than as an agent. Such coordinators vary according to their responsibilities. The master architect (MA) is undoubtedly the coordinator of the overall interests, overall coordinating the interests of all parties, but also balancing the design output of the architects of all regions. The block architect (BA) completes the balance of related interests in the sub-projects and feedback to the overall control of the master architect. The master architect can also revise the established guidelines after the discussion with the block architect.

Another function of an architect is his presence. Compared with planners, architects' profession has a comprehensive and strong professional background since its birth. The well-known professional differentiation is only

建筑师完成职责赋予的相关职能；另一方面还兼具政府、开发商、终端使用者的三重视角，从各自的角度对相关事务进行判断，并主动协调相互之间的利益冲突。因此预期说是代理人身份，不如将其定义为协调者。这种协调者因其职责的不同而不同。总建筑师（MA）无疑是整体利益的协调者，总体统筹各方的利益，也平衡各地块建筑师的设计产出；地块建筑师（BA）在分项项目中完成相关利益的平衡，并反馈于总建筑师的总体控制；总建筑师也可在与地块建筑师的商讨之后，修订既定的方针。

建筑师的另一个作用在于他的在场性。相对于规划师，建筑师职业从其诞生起就具有一个综合性极强的职业背景，耳熟能详的专业分化只是在近现代才出现的职业分工，仅此而已。真正的建筑师应具备“三只眼”：第一只眼看到的是建筑的本体，关于空间、功能、结构等建筑学的核心内容；第二只眼是“广角镜”，能透过建筑个体看到城市的整体架构和内在关系，建筑设计的优劣最终接受城市最公正的评判；第三只眼看到的是“微距镜”，看到的是建筑的营建细节，发现密斯定义的“上帝”，并将其赋予在建筑的“灵魂”之中。正因为此，阿尔伯蒂才将建筑师定义为“全面发展的人”。如果我们将麻磡的改造称之为城市设计，那么城市设计也从诞生之日起，就始终徘徊于建筑师的世界，关键在于我们用什么样的方式将其捕捉、善用。

麻磡试验已经过去了半年之余，借此文回顾在持续 4 个月时间里的相关思考是一件令人享受的事情。欣慰在于通过麻磡试验，我们发现了一条当在大拆大建已成为过去，目前我们还茫然不知所措的情况下，处理复杂的城中村改造或街区改造的可能之路。虽然在此试验中我们屏蔽了一些问题，只强调了需要主要面对的问题方面，但其中主要路线的选择和价值标准的建立仍然具有极强的现实参考性。众所周知的拆迁难、改造难已经不再仅仅是一种技术问题的存在，而是牵一发动全身的社会问题与技术问题的叠加，在寻得了新的方法语言之后，我们有这样的自信。因为我们在场。

参考文献

[1]Yasunori Kitao.Collective Urban Design: Shaping the City as a Collaborative Process [M].Netherlands: DELFT UNIVRTSITY PRESS, 2005.

[2]Matthew Carmona 等 . 城市设计的维度 [M]. 冯江等译 . 南京：江苏科学技术出版社，2005.

[3] 斯皮罗 · 科斯托夫 . 城市的形成 [M]. 单皓译 . 北京：中国建工出版社，2005.

[4] 刘梦琴 , 傅晨 . 城中村国内研究文献评述 [J]. 城市观察，2010（06）：177-185.

[5] 程丹丹 , 吴雪颖 , 孟凡炀 . 深圳城中村更新模式对比——以水围村和大冲村为例，共享与品质 [J].2018 中国城市规划年会论文集（02 城市更新），2018：413-423.

[6] 孙梦水 . 基于复杂系统理论的“城中村”发展研究：以北京为例 [D]. 北京农业大学博士学位论文，2013.12.

[7] 刘蕾 . 城中村自主更新改造研究——以深圳市为例 [D]. 武汉大学博士学位论文：2014.12.

a division of labor in modern times, that's all. A real architect should have "three eyes": the first eye sees the essence of architecture, the core content of architecture about space, function, structure and so on. The second eye is the "wide-angle lens", which can see the overall structure and internal relationship of the city through the individual building, and ultimately accept the advantages and disadvantages of architectural design follow the fair judgement city. The third eye sees the "micro mirror", the construction details of the building, finds the "God" defined by Mies Van der Rohe and endows it with the "soul" of the building. It is for this reason that Alberti defined architects as "all-round people". If we call the renewal of Makan village as urban design, then urban design has been hovering in the world of architects since its birth. The key lies in how we capture and make good use of it.

Half a year has passed since the Makan experiment. It is enjoyable to review the work lasting for four months in Makan village. Thankfully, through the Makan experiment, we found a possible way to deal with the complex issue of urban village and block renewal when the demolition and construction has become a thing of the past and we are still at a loss. Although we shielded some problems in this experiment and only emphasized the main problems we need to face, the choice of main routes and the establishment of value standards still have a strong practical reference. As we all know, the difficulties of demolition and reconstruction are no longer just a technical problem, but the superposition of social problems and technical problems that affect the whole body. After finding a new method language, we have such confidence. Because we are here.

城中村除了私宅以外，村集体股份公司普遍利用空地建设厂房出租。随着产业升级，厂房物业价值降低，另一方面第三产业的增长也面临办公空间场所的需求旺盛。厂房空间大、租金便宜、区位较佳等多方面优势促进了服务业特别是文化、创意产业的进驻。城中村、旧厂房的产业化改造趋势像浪潮一样，在深圳蔓延开来。

Besides dwelling place, public area is also commonly utilized by the collective stock company of village to build factories for rental. Owing to industrial upgrade, the value of factories are going down while a boom in the third-industry is pulling up demand for office space. Meanwhile, advantages such as large space in factories, low rental and good location attract lots of innovative industry to stay in urban villages. The renovation of urban villages and old factories has become a trend in Shenzhen.

2 改造驱动创意

由功能改变方式主导的城市更新研究

RENOVATION MOTIVATED INNOVATION

THE STUDY OF FUNCTION TRANSFORMED AS A TYPE OF URBAN RENEWAL

2.1

既有工业建筑改造与创意社区的辩证法

A DIALECTICAL VIEW ON THE RENOVATION OF EXISTING INDUSTRIAL BUILDINGS AND CREATIVE CLUSTERS

Xiao Jing 肖靖

---

1. 深圳既有工业建筑的历史语境

高密度城市环境中的既有工厂建筑改造，转而以艺术园区或创意产业为主导特色，应当发轫于 20 世纪末的艺术家聚落改造运动。早期位于北京、上海和广州的旧厂房被艺术家占领，因其相对低廉的租金和较为适宜改造的大空间类型，催发出一大批艺术园区的蓬勃兴起。深圳实际上自 20 世纪 80 年代上半期便兴建了一整批工业建筑，在“三来一补”的建设理念下，这个地处边陲的经济特区开始全速完成多达 10 个出口工业区的建设任务。当时主要兴建的是位于罗湖区的八卦岭工业区和南山的蛇口工业区，建设规模较大的其他园区还包括上埗、莲塘、布吉和宝安。招商引资而导致工业厂房的快速修建，诸如入驻蛇口较早的日本三洋电器公司，仅花了不到三个月的时间，便完成了从洽谈、安装生产设备、再到投产的全过程。如今，三洋旧厂房已经被改造成名为“南海意库”的设计研发中心，集中了南山区以创意设计、建筑设计和媒体公司为主的现代办公与娱乐休闲场所。根据 1982 年编制的《深圳经济特区社会经济发展大纲》的相关要求，深圳特区将建设成为“以工业为主，兼营商业、农业、住宅、旅游等多功能综合性经济特区”。以此为基调，20 世纪 80 年代包括华侨城等国有企业和单位，由原先的农业生产发展转向为工业化发展模式，目的在于吸引海外华侨和港澳同胞进入深圳投资建厂。而那时所兴建的加工厂房如今被改造为 OCT 华侨城文化创意园（图 1）。

进入 20 世纪 90 年代以后，深圳再次呈现出新的转向。《深圳市城市总体规划 1996-2010》则确立了以现代产业协调发展为目标的综合性方针，深圳将着力打造经济中心和金融中心，进而转型成为全国性乃至国际化的中心城市。都市精品生活空间逐渐被新型的中产阶级所使用，品味和情趣广受追捧，以消费为主要特色和导向的大型商业综合体和娱乐休闲设施正在成为深圳城市片段中的最重要因素和场景，社会空间生产与商业资本主义意识形态的逐步融合，使得原本功能至上的建筑原型开始向大众消费与文化资本为特征的创意产业需求迈进。无独有偶，近年来国际影响力逐渐扩大的深圳城市 / 建筑双年展，屡次将既有工业建筑厂房及环境作为主要卖点，2005 年华侨城创意园和 2013 年的蛇口大成面粉厂，都是利用旧工业建筑的转型

1.Historical Context of Existing Industrial Buildings in Shenzhen

The Renovation of existing industrial buildings in the high-density urban environment has a long tradition since the last century when the art villages became the driving force of the transformation oriented by art parks and creative clusters. In the early stage, old factories in Beijing, Shanghai and Guangzhou were taken up by artists due to their comparatively lower rent costs and larger spatial orientation. A large number of art parks came to surface. Shenzhen, on the other hand, accumulates a series of industrial parks under the trading policy of "Three Types of Processing Plus Compensation" since the early of the 1980s. There were more than ten industrial parks of this kind under construction during that time, including those at Bagualing in Luohu, Shekou in Nanshan, Shangbu, Liantang, Buji and Bao'an. Foreign capitals rushed in and made possible a high speed of industrial development. For instance, Sanyo Electricity Corporation from Japan established its first manufacturing park in Shekou, finishing all the procedure of negotiation, onsite installation and production within less than three months. Now, the original industrial buildings of Sanyo have been transformed into the high-end Nanhaiyiku R&D Center, a large-scale modern commerce and entertainment cluster with offices for creative design, architectural design and media. According to "Shenzhen Social and Economic Development Plan 1982", the special economic zone of Shenzhen would develop into a modern city targeted on the industry, commerce, agriculture, real estate and tourism. Therefore, those state-owned corporations who focused on agriculture only would turn to industrial businesses and draw more overseas Chinese and compatriots from Hong Kong and Macao to set up local factories in Shenzhen. As one of the achievements during that time, the Overseas Chinese Town (OCT) established many processing factories which now become the OCT cultural park(Fig.1).

After the 1990s, Shenzhen saw something else. The new master plan 1996-2010 claims that the harmonious development of the modern industry is of top priority, and Shenzhen would become the economic and financial centre among those world-leading metropolitan cities. The emerging bourgeoisie widely accepts the elegant taste of the elite urban culture. Large-scale commercial centres and leisure places grow to be part of city-branding under the veil of consumption. Social and spatial production merge with commercial capitalism, which drives architecture from a functional basis towards the mass consumption of cultural capitals. Recently, Shenzhen City Biennale of urbanism and architecture receives more international influences, and it inclines to reuse industrial buildings as the exhibition sites. OCT Cultural Park in 2005 and Shekou Dacheng Flour Mill in 2013 are both the excellent cases in re-using industrial buildings to reflect and regenerate old urban districts. Shenzhen just issued the manifesto of the City Development Outline 2030 – Global City via

图 1 华侨城 OTC 文化创意园
Fig.1. OTC cultural and creative park

来反思和带动老旧城区的再生。深圳凭借《2030 年城市发展策略：建设可持续发展的全球先锋城市》的宣言，在进入存量建筑时代的过程中，努力发展侧重创造性活动和高附加值的产业升级。

2. 文创产业与创意社区

文化产业研究翘楚——英国伯明翰学派认为，进入后工业、高密度城市环境中的物质资源再生与利用成为公众话题，低能耗、可持续、高回报的创意产业将成为带动整个城市发展的新模式。 而文化创意产业研究领域内，对建筑设计从业人员的自身定位有着不同的意见。例如创意产业学者理查德 · 弗罗里达（Richard Florida）认为，包括建筑师、信息工程师、医生等从业人员将成为创意城市环境中的重要成员，而通常我们所认为的演艺类、艺术创作类人员地处外延，并不是定义创意产业的核心力量。 如果因此为出发点，那么城市街区更新后的建成环境将会是以商业办公为主要形式的类型化空间，而不是诸如大芬艺术村或观澜版画村等艺术产业为主的场所语境（尽管这两处均以城中村改造升级为背景），因为太过于“波希米亚式”的城市环境，与现代都市的乡绅化潮流格格不入。

如果从文化经济学者大卫 · 斯罗斯比（David Throsby）的观点看来，恰恰是传统艺术创作活动及其空间生产逻辑才是推动文化产业发展的原初动力。工业化体系中的艺术产品生产模式将大规模地回应中产阶级的消费趣味和生活需求，以往高级艺术形式将被低级趣味消费所取代，精英文化内核被替换为广泛的群众基础，斯罗斯比相信这才是联合国教科文组织推行“文化产业”时的初衷。一个良好的、推动创意产业发展的城市环境，需要以促进社会参与度和友善环境为基础，增强自身品牌效应与艺术活力，并提升整个街区成为“创意社区”的层级。在这个过程中，文化旅游仅仅是低级资本生产阶段，城市与建筑设计如果不能突破追捧伪文化符号和场景的桎梏，那么由此引发的旧城改造、城市更新、文化遗产保护乃至旅游地产开发，都会成为

Sustainable Development, claiming that in this very era of existing buildings, the creative and high-value industrial upgrade has become far more important than ever before.

2.Cultural Industries and Creative Cluster

The Birmingham School claims that the cultural industry in the post-industrial age should attach more importance on regeneration and re-use of material resources in the high-density urban environment. Creative industries with low energy cost, sustainability, high values could be a driving force of new urban progress. Studies of the cultural industry believe that architects play a special role in this procedure. Richard Florida, a leading scholar in this field, says that the key players in a creative urban community that includes architects, information engineers and doctors. On the contrary, those traditional factors like artists would no more be at the core of this new industry. Following this principle, the future of urban regeneration would welcome the building types of commercial and office space, rather than art villages like Dafen Painting Village or Guanlan Print Village which turn out to be too Bohemian and free style, and therefore in contrast with bourgeois gentrification.

From a different point of view provided by Professor David Throsby, it is exactly the traditional art creation and its spatial production that addresses the power of cultural industry. Artworks being systematically manufactured and distributed through industrial lines could reflect the living needs of the middle class. Instead of high art and elite culture, the mass would embrace low art. Throsby believes that, with a deep understanding of this trend, UNESCO promotes cultural industry worldwide. Nevertheless, a good and pro-creation urban environment requires a fundamental acceptance of social participation to maintain the effect of self-branding and artistic vitality. It foresees the emergence of creative clusters on the street/block level. During this process, cultural tourism is at lower levels of capital. The urban and architectural design would be trapped in the circle of pseudo-cultural consumption. Urban regeneration, heritage protection and tourist real estate would be part of this visual feast while compromising in situ cultural identity and urban images.

The regeneration of existing industrial buildings in Shenzhen and Shanghai has been more prudent within this particular context. A large number of spared industrial buildings become the potential subject of art parks with a series of design strategies, including 1) on urban design, to change it from industry-based to public-oriented space, as well as its efficiency of land development and function (OCT Park in Shenzhen and Xintiandi in Shanghai); 2) on architectural design, taking Nanhaiyiku for example, to introduce high-quality space by implementing new and sustainable structures while keeping the integrity of old ones; 3) on environmental setting, to incorporate new manners according to contemporary users' needs in terms of their new life, work and

抹杀在地文化认同、扭曲城市意象的视觉消费盛宴。

包括深圳、上海在内的既有工业建筑改造，就是在这样的整体氛围中开始酝酿、发展和成熟起来的。大量闲置工业厂房建筑走到台面，作为崭新的艺术园区而被重新发掘出价值。落实到具体的改造升级措施层面，不限于以下几个特点：1. 城市设计层面，工业用途转换为公共功能，提高土地使用密度和功能整合效率（如深圳 OCT 创意园、上海新天地）；2. 设计手段方面，更换或回收利用旧有结构、引入绿色环保技术，打造高品质建筑环境（南海意库）；3. 环境认同方面，结合当代城市人群使用方式，提供新型生活、工作与消费模式，成为带动周围片区空间价值的“触媒”（大鹏区满京华美术馆区、上海杨树浦滨江创意园）。但是，在缺乏对工业建筑类型及其价值的真正理解时，既有工业建筑改造很大程度上将变为房地产产业链当中的下游环节，工业遗产沦为文化消费商品的空间再生产手段，成为文化资本的附庸。像上海八号桥这种旧厂房尽管改造时间较早、开发相对成熟，但也在很长一段时间内无法扭转高度依赖旅游和销售周边纪念产品的营销模式，虽然各种世界顶级建筑设计所云集此地设立分公司，但对周边地块（包括田子坊开发）的带动，相比起工业建筑文化旅游来说影响甚微。台湾地区建筑泰斗汉宝德先生在反思文创产业时曾说过：“……这种‘低技’的古迹再利用，并不是文化产业，而是一般房地产之用途而已，不可能创造更大的产值……老实说，这部分的效能在产业上看还是很有限的，甚至可以完全忽略。” 在建筑设计作为工业建筑改造、提升文创产业的手段时，未经辨析地使用现下流行的设计语汇，将使我们无法区分真正意义上的遗产保护与套现政府资助或反向促成商业住宅混合地产模式工具之间的差别。

3. 作为创意产业的城市建筑学

在库哈斯看来“无法阻挡”的世界性资本生产与消费的体系中，无论是城市、建筑还是个人的价值都岌岌可危。传统建筑产品如何变现为大众艺术消费的对象，是否会成为现下创意产业的唯一标准，这仍是个未解的题目。既有工业建筑只是其中一个看上去很美的外衣，如果缺乏应对策略，它将阻碍我们去深入理解城市诸多历史肌理迭代的真正价值，也会隐藏掉其对城市产业升级所带来的真正挑战，仅剩下如同文化旅游学者约翰 · 尤里（John Urry）所说的“欢快的表情”。终究，这种整合过程的焦虑会传导到城市建筑学对自身的定义，从而影响我们在面对都市困境和日常生活时所做出的判断。

leisure styles, and therefore to hybrid a spatial medium for surrounding districts (Manjinghua Art Museum Park in Dapeng, Shenzhen, and Binjiang Creative Park in Yangshupu, Shanghai, etc.). However, without sharp insights upon the value of industrial buildings, the regeneration of this kind would be subject to the cultural capital of real estates. Industrial heritage thus confronts various threats from the spatial reproduction of cultural consumption. Take the Bridge 8 for instance. This project is one of the earliest industrial regeneration in Shanghai and claims to be relatively mature in developing mechanism. Actually, this project, for a long time, could not balance cost and income if without souvenir sales from industrial tourism, even if many world-leading architectural design companies have set up branches here for years. Po-teh Han, the famous architectural critic from Taiwan, maintains that the low-level regeneration of heritage in this manner belongs not to the cultural industry but ordinary real estates without higher values, and honest to say, without values at all in terms of industry upgrade. Architecture could be of no significance if contemporary design language fails to provide identifiable methods for heritage protection. Otherwise, it may deliver negative effects that commercial developers would intrigue to see when they reclaim governmental reimbursement in the name of providing social benefits.

3.Urban Architecture as Creative Industries

To Rem Koolhaas, the urban, architectural and individual values are under attack by inevitable production and consumption of global capital. The question of whether traditional architecture could fit the needs of the mass and become the top criteria of creative industries is still looming. Without specific strategies, the existing industrial buildings would camouflage the authentic values of historical urban context and the challenges during the process of industry upgrade. Cultural Tourist Scholar John Urry once warned readers of this danger as the “tourist gaze”. After all, it may distract the meaning of urban architecture and our judgment over the problems of everyday life.

参考文献

[1] Department for Culture, Media and Sport Creative Industries Mapping Document, Report of DCMS[G]. London: DCMS, 1998.

[2] Richard Florida. The Rise of the Creative Class: And How It’s Transforming Work, Leisure, Community and Everyday Life[M]. New York: Basic Books, 2002.

[3] David Throsby. Economics and Culture[M]. Cambridge: Cambridge University Press, 2001：112-113.

[4] Sharon Zukin. The Culture of Cities[M]. Oxford: Blackwell, 1995：271.

[5] 徐苏斌 . 从文化遗产到创意城市——文化遗产保护体系的外延 [J]. 城市建筑 . 2013.9.

[6] 汉宝德，文化与文创 [M]. 台北：联经出版，106，156.

2.2

建筑与公共空间的自发性建造

SPONTANEOUS CONSTRUCTION OF BUILDINGS AND PUBLIC SPACES

XuKai 许凯

城中村作为城市创新的空间实验场

The village in the city serves as the space experiment field of urban innovation

1. 城中村与创意产业相遇

1912 年，“创意经济”的先驱约瑟夫·熊彼特（Joseph Alois Schumpeter）提出，创新是现代经济发展的根本动力（Hagedoorn J，1996）。2001 年英国政府提出，“所谓创意产业，就是指那些从个人创造力、技能和智力中获取发展原动力，并且可以通过开发知识产权，创造出潜在财富和就业机会的行业”。有学者指出，21 世纪的产业将越来越依赖于以创新为基础的知识生产能力（Landry C，Bianchini F，1995）。今天，创意产业已经被视为经济繁荣不可或缺的部分。

创新产业对城市发展至关重要，但是城市为此准备好了吗?

从欧洲、美国到韩国和日本，那些创意产业聚集发展的地点，往往是出人意料的。瑞士巴塞尔的 Werkraum Warteck（图1、图2），原本是一个老啤酒酿造厂，随着创意人群的不断集聚并对建筑的内部功能进行改造，使原先废弃的工厂公共空间系统得到完善，为各种文化项目提供了丰富的空间场地。现在的酿造厂内，不同的项目、创意企业和创意人群在同一个屋檐下共存，这里也成为城市事件的多发区域，吸引着更多社区居民和游客参与其中，生活模式和空间形式在此不断变革和发展。奥地利维也纳的 Reindorfgasse（图 3），它原本是 15 区的一个红灯区，“卖淫”“犯罪”“肮脏”等消极词汇曾是这个区域的标签。大约在 10 年前，创意人群逐渐进驻这个区域并对其所租赁的空间进行自发性地改造以适应创意产业对空间的需求。如今越来越多的创意活动和人群乐意选择这个老街区，每年在此都会举办露天音乐节、街头集市、社区游学、读书会等活动，创意和生活在这里碰撞。韩国釜山的甘川洞文化村，则呈现出一种“全村创新”的壮观文化图景（图 4）。在国内，北京的 798 艺术区、宋庄艺术区、草场地艺术区，上海的田子坊、M50，深圳的大芬村，厦门的曾厝垵、沙坡尾、顶澳仔猫街（图 5）等案例也是通过对城区空间进行“自下而上”的自发改造形成创意社区，最终达成对城区空间整体自我更新的目的。

图 1 Werkraum Warteck 的创新建筑形式
Fig.1 The innovative architectural form in Werkraum Warteck

图2 Werkraum Warteck 中的公共活动
Fig.2 The Public activitiesin Werkraum Warteck

1. The village in the city meet creative industries

In 1912, Joseph Alois Schumpeter, the pioneer of "creative economy", proposed that innovation is the fundamental driving force of modern economic development (Hagedoorn J, 1996). In 2001, the British government put for-ward that "the so-called creative industries are those industries that derive the driving force of development from individual creativity, skills and intelligence, and can create potential wealth and employment opportunities through the development of intellectual property". Some scholars pointed out that industries in the 21st century will increasingly rely on innovation-based knowledge production capacity (Landry C, Bianchini F, 1995). Today, creative industry has been regarded as an indispensable part of economic prosperity.

Though innovative industries are vital to urban development, are cities prepared for them?

From Europe and the U.S to South Korea and Japan, the places where creative industries cluster are often unexpected. Werkraum Warteck (Fig. 1 and Fig. 2) in Basel, Switzerland, was originally an old beer brewing factory. With the continuous gathering of creative people and the transformation of the internal functions of the building, the previously public space of the abandoned factory has been improved, providing abundant space for various cultural projects. In the current brewery, different projects, creative enterprises and innovative talents are under the same roof, which has also become an area of frequent urban events, attracting more community residents and tourists to participate in it, and the life mode and spatial form are constantly changing and developing. Reindorfgasse (Fig. 3) of Vienna, Austria was originally a red-light district in District 15. Negative words such as "prostitution", "crime" and "dirty" were once labels of this district. About a decade ago, creative people gradually moved into the area and spontaneously adapted their rented space to meet the needs of the creative industry. Nowadays, more and more creative activities and people are willing to choose this old neighborhood. Every year, there are open-air music festivals, street markets, community study tours, book clubs and other activities. Creativity and life collide here. The gangchon dong cultural village in Busan, South Korea, presents a spectacular cultural landscape of "village innovation" (Fig. 4). At home, Beijing 798 art zone, songzhuang art district, caochangdi art district, Tianzifang in Shanghai, M50, Dafen village in Shenzhen , Cengcuoan in Xiamen, ShaPo tail, (Fig. 5) case is based on "bottom-up" spontaneous innovation of the urban space form the creative community, eventually achieve the overall self-renewal in space in urban areas.

图3 Reindorfgasse 中的自发性公众集会
Fig.3 Spontaneous public gathering in Reindorfgasse

图4 釜山甘川洞文化村的文化景观
Fig.4 The cultural landscape of busan ganchuan dong cultural village

It is difficult for creative industries to take place in urban areas where both space and function have been determined. Instead, they prefer the abandoned factories, declined neighborhoods and villages in cities mentioned above, and

图 5 厦门猫街的广场
Fig.5 Cat street square in Xiamen

创意产业很难在空间和功能都已经确定的城区发生，而却偏好上述的废弃工厂、衰败的街区和城中村，采用一种“非正式”的方式来改造房屋和外部空间，最终实现一种群体性创新。原因是什么呢？首先，经济原因是不可回避的，创意产业企业在初创的阶段往往较难支付较高的租金，导致他们只能用这种“非正式”的方式去租用本来并非为产业准备的空间。更重要的是，这些“另类”的城市地点，符合创意产业及其发展的某些特质：

（1）求新、求异的“另类气质”，符合创意产业的文化倾向

（2）较宽松的外部环境，让“非正式建造”得以发生

（3）低层高密度的空间基底，使得高品质公共空间容易实现

城中村是中国城市化过程中的一个特殊现象，在南方城市中非常普遍。很多城中村成为城市里非常独特的存在：密度奇高、人群混居。在很多官员和管理人员眼中这些城中村是一些脏、乱、差的所在，恨不能清扫之而后快，更何况这些城中村的用地往往价值连城。在进城务工的新市民或那些渴望成为市民的人眼中，这里是较容易落脚的住所：离上班的地方近，价格便宜。城中村也成为他们中间一些人创业的起点。艺术家、建筑师和文化工作者则在城中村里看到一种久违的气质，既陌生又似曾相识，充满了烟火味，也充满了机遇，和周边那些昂贵的光鲜的楼盘们迥异其趣。

城中村竟然成为创意产业生根发芽的理想地！在深圳大芬村，村民的住宅成了画工工作、住宿的地方，离村子中心比较近的那些房子，都被开放成画廊，村民也乐见其成地做起房屋出租的生意。其他产业、博物馆、绘画材料、艺术培训也遍地开花。在厦门的曾厝垵，村民住宅成了特色的民宿酒店，村子成了旅游目的地。城中村和创意产业的相遇，迸发出极大的活力，特色的城区诞生了，令人欢欣鼓舞。

有意思的是，作为专业的城市规划和正式的建筑设计，缺席整个过程，这让这里发生的一切，包括房屋改造、新建和室外空间的优化，都带有“野生”的色彩。然而从结果上看，这个过程显然是有积极意义的：城区特色得以形成、产业欣欣向荣、百姓安居乐业。对这些“非正式”的建造行为如何评价；涉及建造本身，他们的机制和表征是什么，是需要我们去思考研究的。

图6 深圳大芬村的城市肌理
Fig.6 The urban texture of Dafen village, shenzhen

2. 从建筑改造到共同营造的公共空间：回到一种久违的形式

我们对自发性营造的考察，来自于两个城中村案例：厦门的曾厝垵和深圳的大芬村（图6、图7）。它们都是通过创新产业的发展，由外来的创意人群和村民一起，对建筑和空间进行积极的营造，形成今天这样的局面：原有的城市肌理得以保留、改建的建筑富有特色，公共空间充分发育。我们将从需求、营造的历程、空间演变、建筑等多方面来进行介绍。

图7 厦门曾厝垵的城市肌理
Fig.7 The urban texture of Zeng cuoan, xiamen

1）创意产人群和企业的内在需求

创意企业和人群，作为自发性建造的主体，他们所需求的空间是什么样的？

首先，自发性建造往往不是从零开始的，而是以租用房屋为基础进行改造而实现的。作为改造的基础的原始建筑应有小型化、灵活度

adopt an "informal" approach to transform houses and external spaces, so as to finally realize a kind of collective innovation. Why? First of all, economic reasons are inevitable. It is often difficult for creative industry enterprises to pay high rent in the initial stage, so they have to use this "informal" way to rent space that is not originally prepared for the industry. More importantly, these "alternative" urban locations fit certain characteristics of creative industries and their development:

(1) The "alternative temperament" of seeking novelty and difference conforms to the cultural tendency of the creative industry
(2) A looser external environment allows "informal construction" to take place
(3) In low-rise high-density space base, high-quality public space is easy to achieve

Urban village is a special phenomenon in the process of urbanization common in southern cities of China. Many urban villages have become very unique in their cities for their extremely high density and mixed population. In the eyes of many officials and managers, these villages are dirty, messy and poor, and should be demolished as soon as possible. Moreover, the land in these villages is in high value. To the new migrants or would-be citizens, it is an convenient place to live with comparatively shorter commuting time and lower rental. Urban villages are also a dream incubator for migrants. Artists, architects and cultural workers feel an air, strange but familiar, in the villages, wordly and full of opportunities, which is different from the expensive and shiny buildings around them.

Village in the city has become an ideal place for creative industry to root and sprout. Dafen village, Shenzhen, villagers' houses have become places for artists to work and live. Those houses near the center of the village have been used as galleries, and villagers are happy to rent out their houses. Other industries, museums, shops for painting materials, and art training institutions have blossomed. In Zengcuo An, an urban village in Xiamen, villagers' houses became B&B hotels, promoting the village into a tourist destination. The encounter between the urban village and the creative industry bursts out great vitality.

Interestingly, the absence of professional urban planning and formal architectural design from the whole process lead to an disordered development pattern here. House renovations, new construction and optimization of outdoor space etc. are all in a wild color. However, from the result, this process is obviously positive: urban characteristics are formed, industries are flourishing, and people live and work in peace and contentment. How to evaluate these "informal" construction behaviors; To construction itself, what are its mechanisms and representations, is something we need to think about.

高的特点。创意企业对小型化的工作空间有着持续的、巨大的兴趣。这与企业的规模有很大的关系。在全球化的趋势下，产业细分越发精细化，企业规模就越小。这些小型的企业往往在全球化的产业链中占据一环，依靠创新来维持它的活力和竞争力，而不必去按照传统方式做大做全产业链以寻求规模扩展。它们在创业的初期往往无法负担很大的办公空间，且对空间的独立性有很高的需求。它们非常在意空间的灵活度，因为企业的规模随时会因为业绩和发展思路的不同而变化，空间则需要对这些变化进行适应。因此，城中村就因为能够满足这方面的需要成为首选。

第二，对于很多创意企业来说，工作空间是展示他们产品的重要场所（如产品设计、时装设计等产业），空间是它们品牌形象的一部分（如文化、媒体或者设计类型的企业）。这会驱使他们首先选择有空间特色的城区来入驻，并在建筑的内部或外部空间改造方面投入很多的精力和创新力。“开放性”是空间改造重要方向，例如，大芬村和曾厝垵的沿街建筑在底层的外墙都会被打开，用于展示绘画作品或者其他用以售卖的物品。

第三，公共空间是创意企业和人群共同的需求。对各自房屋的独立改建行为，最后都会演变成群体性对公共空间的改造，从而推进社区的整体提升，这是创意社区最令人振奋和赞叹之处。生产新的知识，是创意产业的基本职能。新知识的生产往往是来自于不同产业个体（企业或个人）之间的相互联系和合作，它需要依赖的一定的物质空间，这就是公共空间。也只有公共空间的存在，才让创意社区作为“社区”的特质真正得以形成。大芬村和曾厝垵的街巷，都是对公共空间进行“共同营造”的结果。

2）从建筑改造到公共空间营造

从个体建筑改造到群体空间营造，经历一个过程。首先，个体企业租用房屋并置换房屋的功能，进而对房屋按照他们的需求进行内部改造。接下去，一些企业开始寻求更进一步的发展，这包括向相邻的单元进行扩张并将那些本来分散的空间进行连接。他们会在可能情况下对建筑的立面和形体进行更改来适合企业的功能使用，例如建楼层。很多的创意企业对公共空间会有一定的需求，用以增强与游客（很可能会成为企业的客户）的交流，这会驱使企业对建筑物的沿街面进行改造（“破墙开店”是其中最典型的），并改造和利用与自己房屋接壤的外部空间地面，如增加地面铺装和增设桌椅外摆。企业的数量增加并彼此相连时，大家就会发现将各自的外部空间整合起来规划是对大家都有利的，于是街道就产生了。

当然，实际情况可能更复杂一些，也并不是所有的改造诉求都能够成功。除了创意企业和创意人员，仍旧居住在区内的居民或区内既有功能的使用者也构成利益相关者的一部分。他们对空间当然有不同的需求。例如，有些居民可能就对喧闹而不那么安全的公共空间感到十分反感，而其他一些居民却仍可能支持创意企业对外部空间的改造，因为这些居民作为出租房屋的业主在整个城区的发展中获益。居民和企业之间、居民和居民之间、企业和企业之间的那些联合行动或斗智斗勇，正推动着创意社区一次次的空间转变。

因为上述的发展机制，导致公共空间的发展是非常具有生长性

2.From architectural transformation to co-creation of public space: a return to a long-lost form

We make an research in spontanous development, from the caces of two villages Cengcuoan Village and Dafen village (Fig. 6,Fig. 7). Through the development of innovative industries, creative migrants and villagers create buildings and spaces together, and form the present situation: the original urban fabric, the characteristic buildings and the fully developed public space. We will introduce the demand, the process of construction, the evolution of space and architecture.

1) Internal needs of creative production crowd and enterprises

What kind of space do innovative enterprises and people, as the main body of spontaneous contruction, need?

First, spontaneous construction is often not done from scratch, but rather on the basis of a rented house. The original building as the foundation of the transformation should be small and flexible ro reconstruct. Innovative companies have continuous and huge interest in miniaturized workspaces. It is related to the scale of the business. Under the trend of globalization, the industry segmentation is more and more refined, and the scale of the enterprise is smaller. These small enterprises often occupy a section in the global industrial chain, and rely on innovation to maintain its vitality and competitiveness, rather than follow the traditional way to make a big industry chain seeking scale expansion. They are often unable to afford large office space in the early stage of entrepreneurship, and have a high demand for independent space. They care very much about the flexibility of space, because the scale of the enterprise can change at any time depending on the performance and development concepts, and the space needs to be adaptive. Therefore, urban villages become the first choice because they can meet the requirements above.

Second, for many creative enterprises, the work space is an important place to display their products (such as product design, fashion design, etc.), and the space presents an image of their brands (such as culture, media or design type enterprises). This will drive them to choose the area with unique spacial characteristic, and invest in the transformation of the internal or external space of the building. The openess of space is an important target for space transformation, for example, in Dafen Village and Cengcuo An Village, walls of the ground floor along the street will be open to display paintings or items on sale.

Third, public space is the common demand of innovative enterprises and people. The independent reconstruction of each house will eventually contribute to the transformation of public space, promoting the overall improvement

的，它往往聚集于一两个点状或线状要素，再慢慢扩张。在量的案例中，上述趋势都比较明显。此外，案例的各阶段发展都呈现一定结构性特征，但这些特征在发展过程中会发生改变。例如，大芬村的发展一开始是沿外围道路，但发展的后期则以主街为中心。曾厝垵的空间结构则一直是沿着主要的街道发展，从点状发展成线状，到最后形成网络结构。（图 8）

图 8 曾厝垵和大芬村的公共空间演变过程
Fig.8 The evolution of public space in Dafen village and Zengcuoan

3）特色建筑类型

在城中村中，建筑类型是比较明确的，即“独立式小型建筑”。一家或者多家企业合用一个建筑单体的现象比较多，产业也可以和既有的功能合用建筑单元。建筑与公共空间的依存关系很明显，也比较容易形成有活力的街道空间。典型的“独立式小型建筑”是大芬村和曾厝垵的村民住宅。在大芬村里，沿着主街和广场的村民住宅，功能都被重新利用以适应油画产业的需求，例如下面的楼层改造为画廊，面向商业街开门和进行橱窗展示。上面的楼层则有多样的功能组合可能，如油画作坊、画具相关产品的车间、画工宿舍等。在曾厝垵里，村民住宅对酒店功能的极大兼容性，正是精品酒店产业在曾厝垵蓬勃发展的原因。基本的改造方式是，村民把围绕宅基地的院墙打开，把房屋以外的自用地开放，对之进行简单环境改善以吸引游客停留。底层改造为商店和酒店大堂，上面的楼层改造成客房，留一部分面积供房屋主人居住。顶层和底层常常都会有加建，以扩大经营和居住面积，尽管该行为是被官方禁止的。

因为建筑都是独立式的，进行顶部加层的可能性较大，向周边进行加建的可能性则较小。建筑的功能改造一般不会影响建筑的基本形态，存在对内部掏空进行空间重组的可能性。立面改造的可能性很大，设计的空间也很大。此类型建筑对外部公共空间有一定贡献，并可能连接成连续的公共空间，如大芬村中的街巷。在曾厝垵中，在保留院墙的基础上，建筑的前区空间变成服务酒店的庭院，对城市而言是一

of the community, which is the most exciting and impressive part in the upgrating process to a creative community. Creating new knowledge is the basic function of the creative industry. The creation of new knowledge often comes from the Interaction and cooperation between individuals (enterprises or individuals) in different industries. It needs to rely on certain public space. Only the existence of public space can realize the characteristics of a community. The image of streets in Dafen village and Cengcuoan village comes from the joint-efforts of the renovation of every single building.

2) Transformation from individual buildings to public space

The transformation from individual building to public space goes through a certain process. First, individual enterprises replace the original functions in rented buildings with functions that fulfill their needs. Then, some enterprises begin to seek for further development, such as expansion into adjacent units and connecting the previously scattered spaces. Where possible, they make changes to the facade and form of the building to fit the functional use of the enterprise, such as adding floors. Many creative enterprises of public space will have certain needs, to strengthen and the communication with tourists (is likely to become the enterprise customers) , this will drive enterprises to modify the interface along the street (tearing down the walls or widening openings for shops is an typical example), and utilize the external adjacent space, extending paving of the ground floor and adding tables and chairs. As more enterprises gather closely in the village, people will find that it is in everyone's business to integrate their respective external spaces, and the street is formed.

Of course, the reality may be a little more complicated, and not all the transformation appeals can succeed. In addition to creative enterprises and creative personnel, residents living in the district or users of existing functions in the district also constitute part of the stakeholders. They certainly have different needs for space. For example, some residents may find noisy and unsafe public spaces distasteful, while others may still support creative enterprises to transform the exterior space, because the latter benefits from the development of the whole village as rent collectors. It is these joint actions or battles between residents and businesses, among residents, and among businesses that drive the spatial transformations of innovative communities.

Because of the development mechanism mentioned, the development of public space starts from one or two points or line elements and keeps expanding gradually. This tendency is especially obvious in quantity. In addition, each stage of the development presents certain pattern, but this pattern change in the development process. The development of Dafen village, for example, started from a pattern that originated from an outlying road, while later it switched to another pattern centred on the mainstreet. Comparatively, the pattern of Cengcuoan village remained focusing on the main street,. originating in units, and from point to line, forming a network

种有特色的半公共空间。（图 9，图 10）

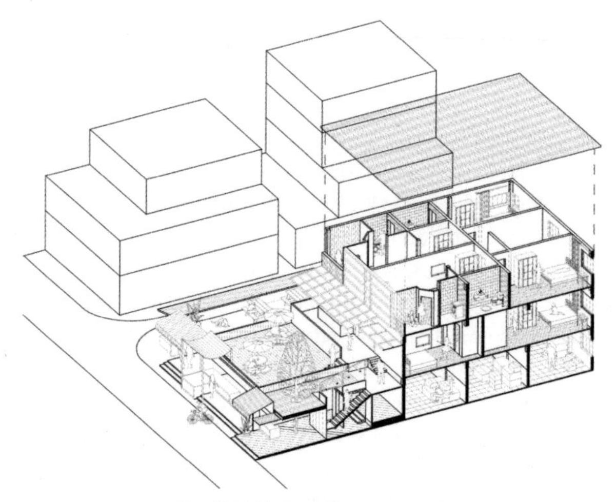

图 9 曾厝垵中典型建筑类型的改造方式
Fig.9 The Transformation of typical building types in Zengcuoan

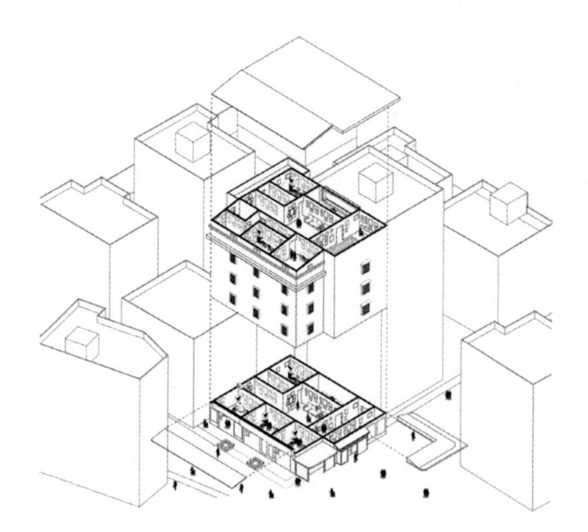

图 10 曾厝垵中典型建筑类型的改造方式
Fig.10 The Transformation of typical building types in Dafen village

structure. (Fig 8)

3) Special building type

In urban villages, most buildings are self-detached and low-rised. It is more common for enterprises to share a single building. Industry can also share building units with existing functions. The interdependence between the building and the public space is obvious, and it is easier to form a dynamic street space. In Dafen village, villagers' buildings along the main street and square have been re-used to meet the needs of the oil-painting industry, such as transforming the lower floors into galleries, creating more openings to the commercial street and promoting window displays. The upper floor provides space for functions such as oil painting workshops, painting tools workshops, dormitories. In Cengcuoan village, the original buildings are compatible for hotels, which prospers later on in the village. The basic way of renovation is to break the courtyard wall around the house site, and open the self-use land outside the house and enhance environment to attract tourists. The ground floor was converted into shops and hotel lobbies, while the upper floor was converted into guest rooms. Additionally, a portion of the space is kept for the owner. Extensions on the top and the bottom of the buildings are common to extend interior space, though banned.

Because the buildings are all self-detached, it is more likely to add the top layer. The functional transformation of the building generally does not affect the basic looking of the building. There is also an opportunity for facade re-design. This type of building contributes to the external public space and may be combined into a continuous public space, such as the streets in Dafen village. In Cengcuo village, on the basis of retained wall, the area in front of a building was transformed into the courtyard of a hotel, creating a semi-public space in the neighborhood. (Fig 9, Fig 10)

3.Questions and Suggestions on Makan village

The theme selected by the urban design camp of the four schools in 2018, the renovation design of Makan village, is a challenging one. The research group pointed out that the development mode of urban villages is no longer the one to demolish and reconstruct, but an alternative one based on the status quo, protecting the interets of residents, preserving local industries and space. This gets rid of the unsustainable mode of pursuing short-term profit and returns to the basic laws of urban development. Combined with the research above on spontaneous construction and urban space construction, we make the following proposals:

3. 麻磡村的问题和建议

2018 年中奥四校城市设计营所选取的题目，麻磡村的改造设计，是一个很有挑战的题目。课题组提出，城中村的发展模式不再是推倒重来，而必须以现状为基础，保护居民、产业和空间基础进行提升发展。这摆脱了以追求短期利益产出为目标的不可持续模式，回到城市发展的基本规律上来。结合上面我们对自发性建造和城市空间营造的研究，我们提出如下的一些思考：

1）产业转型是麻磡村发展的大背景，核心产业的策划是区域发展的前提。制造业衰败后闲置下来的大量厂房建筑，正是未来新兴产业的理想聚集地。离市区较近的距离，周边比较丰富的大专院校资源，都暗示着服务年轻人、创办者和科技创新的产业可以在这里得到很好的生存和发展。

2）村民首先需要“乐业”，才可能“安居”。村民从事什么行业？房屋租赁、零售商业和其他服务业，都是需要产业充分发展作为基础。这进一步确说明了产业发展的重要性。此外，哪里是鼓励产业嵌入的，哪里是主要维持村民生活的，需要进行区分。

3）建筑改造需要考虑建筑的类型和未来的发展方向。麻磡村的“独立式小型建筑”（村民住宅建筑）和“独立式中大型建筑”（多层厂房建筑）是两种具有特色的类型，数量非常多。它们能容纳的产业类型、所涉及的改造方式、激活外部空间的能力是不同的，这是做设计之前需要充分考虑的。

4）政策支持、导则规定下的“自发性建造”，是麻磡村可以采用的一种方式，它可能带来居民参与的积极性、产业发展的多样性和活力，以及建筑类型和公共空间的极大丰富和创新。但是，基础设施、分区等这些宏观结构的树立，必须仍是由“自上而下”的规划来实现。探索一条“自上而下”规划和“自下而上”更新相结合的道路，不仅是对麻磡村，而且对于各地的内城更新项目，都非常重要。

1) Industrial transformation is the background of the development of Makan village, and the planning of core industries is the premise of regional development. The large number of factory buildings idle after the decline of manufacturing industry are the ideal gathering place for the emerging industry in the future. The close proximity to the city and the abundant resources of colleges and universities all indicate that industries serving the young, set-ups and technological innovation can survive and develop well here.

2) Industrial development is the basis for residence in the village. Only developed industries can provide stable income for the residents, promoting them a better life. Also, the functions of industry and residence should be clearly seperated.

3) The type of buildings and future development should be considered in the transformation of the building.In Makan village there are two distinctive types, small-scale and middle-scale. Designers should give full consideration to the scale because their capacibility for interior transformation and activation of external space differ.

4) The "spontaneous construction" under the policy support and guidelines is a way that Makan village can adopt, which may bring the enthusiasm of residents to participate, the diversity and vitality of industrial development, as well as the great richness and innovation of building types and public space. However, the establishment of these macro structures, such as infrastructure and zoning, must still be realized by a "top-down" planning. Besides, exploring a way to combine "top-down" planning with "bottom-up" renewal is very important not only for Makan village, but also for city renewal projects in various regions.

4. Conclusion and vision

This paper studies a very marginal construction behavior. The study is not only for a form of self-organization of urban space and function, but also for

4. 结论和展望

论文研究的是一种非常边缘的建造行为。研究它的原因不仅在于，我们认为它是城市空间和功能自组织的一种形式，能够推动对既定规划的某种修正和补充；更在于我们发现，在与创意产业结合的情况下，自发建造迸发出很强的生命力，成为设计创新和城市空间营造的战场。研究发现，这种带有“野生”特质的建造行为，是由创意产业追求符合自己特点的空间形式和文化符号的需求推动的，它不仅推动房屋自身的改造，还必然会走向房屋之间的相互连接或重组、打破界限，并最终发展为对公共空间的共同营造。从建筑走向社区，从个体建造走向共同营造，这是非正式建造的必经历程。

城中村是自发性建造行为发生非常频繁的地方，自发性建造行为为城中村带来具有创新的建筑和空间，也让城中村变成产业发展的热土。结合本次中奥四校夏令营，我们探讨和模拟自发性建造在麻磡村的可能性。同时我们还提出，自上而下的规划仍是必要的，它能够帮助城市梳理宏观结构。但是，它必须为利益相关者参与城市发展进程提供必要的空间和自由度。

the vatality bursts from self-motivated renovation integrating the innovative industries, creating a place for innovation and city-space construction. The study shows that the "wild" construction is driven by innovative industries in pursuit of space form in accordance with their characteristics and their demands of cultural symbols. It not only promotes the renovation of the buildings, but also inevitably leads to the interconnection or reorganization of buildings to break the boundaries, finally forms public space with joint effort. From architecture to community, from individual construction to co-construction, this is the only course for spontaneous construction.

Urban village is a place where spontaneous construction happens very frequently. Spontaneous construction brings innovated architecture and space to urban villages and makes villages become a popular place for industries. Through this summer camp, we explore and simulate the possibility of spontaneous construction in Makan village. At the same time, we also suggest that top-down planning is still necessary to help the government sort out the macro structure of the city. However, it must provide necessary space and freedom for stakeholders to participate in the process of urban development.

城市更新不仅体现在城市环境的提升，还能使得城市功能与时俱进促进产业发展。城市更新对城市，特别对城市中最早开发的中心城区维持城市活力影响深远。正因如此，城市更新的方式与策略需要在城市规划层面整体统筹，才能合理、科学地引导城市街区发展、转型。

In the process of refurbishment, besides environmental enhancement, facilities in city are also upgraded to catch up with the development of industries. The refurbishment has a lasting influence on the city, especially the vitality of the early zones. Therefore, it is necessary to have an overall planning at an urban level to guide the transformation of city blocks appropriately and scientifically.

3 城市更新的价值

城市更新作为推动城市发展的手段

THE VALUE OF URBAN RENEWAL

URBAN RENEWAL AS A FORM OF URBAN DEVELOPMENT

3.1

城市更新与创新

BRIDGING URBAN RENEWAL AND INNOVATION

Li Lingyue 李凌月

---

1. 城市更新：动力、类型与治理模式

“城市更新”（Urban renewal）特指美国《1949 年住房法》通过后开始的一系列城市中心区的再开发和更新运动。其原因可追溯到 20 世纪 20、30 年代。当时，随着科学技术的进步，美国的新兴产业开始萌芽，并在城市中不断寻求发展空间，而相对集中在内城的传统制造业则逐渐衰落并纷纷外迁，导致内城日益衰败。城市产业结构的这一变化，对城市的空间布局及经济发展产生了深远影响，其结果是传统的工业、低层次和小规模的零售业及破旧的住宅或自行消失，或在内城中被新兴产业替换[1]。（图 1、图 2）

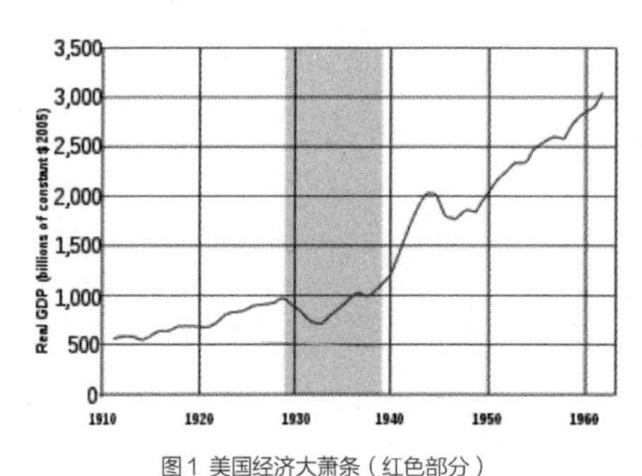

图 1 美国经济大萧条（红色部分）

Fig.1 The great depression (red) (source: https://www.sott.net)

图 2 美国经济大萧条中失业人口在街头找工作，1929

Fig.2 In the great depression, the unemployed were looking for work on the streets，1929，(source: https://www.sott.net)

由此可见，城市更新的动力根植于产业更替和由此引发的城市空间重构，其主要动力包括技术进步，社会经济转型，土地价值提升和社会需求变化。“城市更新”可被定义为“以一种综合性、整体性的观念和行为解决各种类型的城市问题，目的在于对处于变化中的城市地区的经济、社会、物质环境各方面做出长远、持续性的改善和提高”。从空间功能的视角看来，城市更新包含了旧工业区、旧住宅区、旧商业区以及城中村等地区的转型，通过空间再商品化提升地区价值[2]。从交换价值的视角来看，旧商业区的更新常处于两难的境地。一方面她体现了城市身份且已经拥有较高的经济收益，另一方面却存在空间

1.Urban renewal: impetus, types and governance mode

Urban renewal brought into force against the policy background of the "1949 Housing Act". Back to 1920s and 30s, the movement of seeking utopia in the peripheral suburb by the middle income and industry brought remarkable shrinkage of land development market in city. New industry emerged and sought space to settle whereas traditional manufacturing declined and was squeezed out. The result was that low value added, small scale and outdated industries such as crafting and retail gradually disappeared or being replaced. People feared that downtown property value would plummet and their businesses would suffer[1]. The macro context suggests that the determinant motivations of urban renewal rooted in technology progress, socioeconomic forces,increasing land value, and changes in social needs. （Fig.1、Fig.2）

In this regard, urban renewal can be commonly defined as "a comprehensive and integrated response to the diverse opportunities and challenges" presented by urban degradation in particular urban places, the aim is to promote long-term and sustained improvement of urban development in physical, social, environmental, and economic aspects".

From a space functional perspective, types of urban renewal include but not limited to the transition of old industrial, residential, commercial spaces and urban villages wherein space is (re)commodified to upheaval area value[2]. From a perspective of exchange value, renewal of old downtown commercial space often presents a dilemma: downtowns showcasing urban identity already receive considerable economic returns, yet aging and safety problems devalue the space simultaneously. Civic and business leaders regard downtown as the key but troubled ingredient of the overall metropolitan fabric[3]. Displacement of lower-income residents is often intertwined with irreversible gentrification, meaning that they are no longer able to afford their living in current neighbourhood. （Fig.3）Neoliberal large-scale projects upheaved land rent and housing price are not for them but others[4]. Rundown or obsolete industrial cities and regions might profit from culture-based development strategies but is likely to be invaded by capital such that artists or creative men may be forced out under capital pouring[5]. Central to renewal of urban village is the equity in the redistribution of interest derived from land appreciation among main stakeholders. Institutional dichotomy of the rural and urban system is not only the root of the emergence and proliferation of urban villages, but also becomes the obstacle for their regenerations[6]. Essential to all types of urban renewal is the capitalist (re)production of extant urban space[5].

Urban renewal is often featured by "the eradication of blighted areas and the provision of decent homes and suitable living conditions", while the process of urban regeneration is characterized by a more comprehensive and incorporative improvement of urban life[7]. I do not, however, intend to distinguish the

老化和安全性等问题。要保持繁荣，旧商业区就需要为保证足够和持续的人流和光顾率而不得不面对这一难题。市民社会和商业领袖都将商业区更新视为一项重大挑战[3]。旧住宅区的更新往往伴随不可逆的地区士绅化（Gentrification）（图3），原本集聚低收入人士的地区在重建后由于地价和租金上涨和中高收入人士的迁入而不得不迁离到更偏远或条件更差的地区来维持生活[4]。旧工业区的更新多面临资本循环逻辑下文化价值遭受资本侵蚀的风险，衰败或过时的工业空间因艺术文化的注入而再现生机，但复兴的旧区同时也易成为资本逐利的对象而挤压文化艺术工作者的生存空间[5]。城中村更新的关键在于土地增值后利益再分配的公平性，城乡二元结构下的制度二元不仅是出现城中村的根本原因之一，也是其更新的障碍[6]。无论何种类型的更新，其本质都是资本对于城市现存空间的再生产[5]。

图3 毕尔巴鄂城市更新中零售业的士绅化（来源：[7]）
Fig.3 Gentrification of retail in bilbao's urban renewal (source:[7])

美国城市更新运动经历了不同的阶段，含义也随之变化，如城市复兴（Urban Revitalization）、城市振兴（Urban Regeneration）等[8]。含义的区分并非本文重点，演变中城市更新治理模式的变化则具有重要启示。总体而言，西方城市更新治理模式经历了从20世纪70年代政府主导、具有福利主义色彩的内城更新到20世纪80年代市场主导、公私伙伴关系为特色的城市更新，转向20世纪90年代以后以政府、市场、社会三方伙伴关系为导向的多目标综合性城市更新[9]。建立基于空间利益主体的多方参与城市更新机制是达成综合目标实现的重要途径。在新的治理模式下，政府（或公有部门）、私有部门以及社会公众三个主要利益集团在城市更新决策过程中的位置、作用及其相互利益关系，以及由此而产生的城市更新决策实施机制使得该模式的优势充分凸显出来。

2. 创新的城市性与城市更新的创新转向

创新是一个多尺度的空间过程，包含了区域创新体系、都市圈创新活动单元、高科技产业园区、创新产业集群、创意产业集聚区和众创空间。随着后危机时代创新企业和研发人员越来越青睐于在设施完备、交通便利的地区工作和生活，创新的“城市性”越发凸显并得到

meaning between them. Instead, I argue that the governance mode of urban renewal is likely to produce far more profound implications. In all, governance mode of urban renewal evolved from state-led, welfare-based inner city renewal in 1970s, market-oriented, public-private partnership-based renewal in 1980s, to a state-market-society partnership, multi-purpose renewal in 1990s[8]. Building a multi-participating renewal model based on diverse spatial interests is an important path to achieve comprehensive objective. Under this new model, the position, function and interaction of state, market and society have played an increasingly important role in urban renewal.

2.Urbanity of innovation and innovitization turn of urban renewal

Innovation represents a multiscalar spatial process encompassing regional innovation systems, metropolitan innovation units, high-tech development zones, innovation industrial clusters, innovation industrial clusters and hackerspaces. In post-crisis era, innovation firms and R&D talents are increasingly favoring working and living in districts with convenient facilities and transportation, urbanity of innovation becomes prominent and received more attention. Innovation involves professional innovation and technology innovation, the latter is often the key to promoting transformative growth and influences distribution and redistribution of spatial resources[9]. On one hand, spatial distribution of technology and innovation resources is path-dependent, because innovation-based accumulation is continuously self-reinforcing. Moreover, as innovation activities are often more concentrated than other types of production activities, innovation clusters emerge as time went by. On the other hand, innovation difference leads to knowledge spillover that drives innovation development in proximate regions[10].

In addition, alongside urban sprawl, S&T parks once located in suburb have become part of the city center. Traditional high-tech industrial space is experiencing restructuring against the backdrop of infill development. Numerous examples can illustrate the interaction of innovation and urban renewal. For instance, in the transition of industrial land use in central city, firms spontaneously relocate their manufacturing activities, replacing with R&D functions; in mini regeneration of some old residential areas, bottom-up grassroots governance approaches are applied to open up new space for community entrepreneurship.

Innovation involves creative destruction[11]. Technology progress enables capital permeate space production and induces multiscalar spatial restructuring. In the post-crisis era, innovation has replaced advanced manufacturing and high-end service industry, and becomes the protagonist of economic transfor-

更多关注。创新包括专业型创新和科技型创新，后者往往是促进经济变革性增长的关键[10]，深刻影响空间资源的分配与再分配。一方面，科创资源分布有明显的路径依赖特征，因为创新具有不断累积自我强化的特点。又由于创新活动比一般生产性活动分布更为集中，随时间推移则会在城市特定板块呈现出高度集聚态势；另一方面创新“位势差”会形成空间溢出，带动相邻地区的创新发展，从而在城市空间中不断扩散[11]。此外，随着城市的不断扩张，城市中的科创园在区位上大多从市郊变为了城市中心区的组成部分，传统的城市创新空间被动成为存量规划背景下的空间重构对象。在这一背景下，无论是正在形成还是既有的创新空间，都与城市更新在多个面向发生联系。比如，在一些大城市中心城区的存量工业用地，通过企业自发参与和政府引导，将传统的生产制造活动搬离，置换进科技研发功能，成为事实科创产业用地（图 4）；而一些老旧的住区也通过基层政府或社区自治的方式在进行微更新，通过吸引创新人才或开辟创新创业空间激发社区活力。（图 5）

图 4 东伦敦更新与科技城（ Silicon Roundabout ）

Fig.4 East London renewal and technology city （ Silicon Roundabout ）
(source: Flickr, Property UK)

图 5 纽约曼哈顿硅巷创新企业孵化分布（ Sillicon Alley ）

Fig.5 New York Manhattan alley innovation incubator distribution （ Sillicon Alley ）
(source: Digital.NYC, part of the New York City Economic Development Corporation)

mation and upgrading. (Fig.4, Fig.5) In different types of urban renewal, growth machine formed by government and profit-seeking firms is likely to produce innovation space to foster new economy[12]. The innovation turn of urban renewal has been observed in various places and the innovation-driven can be called as innovitization to indicate a new type of urban renewal. Extant research suggests that production of innovation space involves central and local governments, manufactures, firms, universities, and urban residents. From the perspective of government, Buck and While examine initiatives by the UK national government to facilitate urban technological innovation through a range of strategies. They argue for attention to weakened capacity of urban governments to control their infrastructural destiny and the constraints on the ability of the public and private sectors to innovate[13].From the perspective of market, Lindtner draws from long-term ethnographic research on production of China's first hackerspace to reveal how the ideals held by DIY makers, such as openness, peer production, and individual empowerment, are formulated in relation to China's project of building a creative society and economy. It contributes to our understanding of the relationship between technology use, production, society, activism and the state[14]. Using Norwegian Community Innovation Survey data, Herstad demonstrates how firms in the Capital are less prone to engage in innovation activities, but once engaged are more likely to commit strongly innovation inclination than comparable firms located elsewhere[15].

From the perspective of society, Jones explores the attempt to build a nanotechnology sector in Edmonton wherein citizen engagement and urban touring collaboratively contribute to the attainment of entrepreneurial city. Key to avoid homogenisation of urban innovation strategy is the appreciation of those 'virtues of place'[16]. From the perspective of actor partnership, Huggins and Prokop explores the structure of knowledge networks stemming from universities-firms ties. They found that actors in innovative and economically developed regions are more likely to be placed in influential positions within knowledge network architectures[17]. Despite these affluent readings on how different actors involve in innovation city creation, there still leaves room to understand how they manage to foster innovation space (re)production against urban renewal.

3.Bridging urban renewal and innovation

Innovation plays an important role in promoting technological progress. Under the growth-oriented urban development discourse system, the main idea of innovative space production is still to improve production efficiency to enhance the added value of land and gain economic benefits through land appreciation. Therefore, the key driving force connecting urban renewal and innovative space production lies in technological progress and land appreciation: technological progress brings new types of activities and makes new demands on urban space functions, while land appreciation is the intrinsic motivation for

创新是一种创造性破坏[12]。科技进步让资本增值渗透进空间生产领域，从而引发多尺度的空间重构。后危机时代，创新已取代先进制造业和高端服务业，成为经济转型升级的话语主角。在多类型的城市更新中，政府与企业结成的增长联盟更倾向于打造创新空间以培育新经济引擎，而非仅仅依赖创造休闲娱乐空间这一传统手段刺激消费[13]。城市更新的这一转向已在各地的实践中得到开展并有望成为一种新型的，以创新为驱动的复兴模式（Innovitization）。已有研究表明，创新空间生产涉及到国家及地方政府、制造商、企业、大学、城市居民等。从政府的视角，巴克（Buck）和怀尔（While）对英国国家政府一系列创新举措进行了考察，并指出城市政府控制基础设施的能力被削弱，公私部门的创新能力受到限制[14]。从市场的视角，林特纳。（Lindtner）。通过长期民族志研究，探讨了 DIY（do-it-yourself）制造商如何通过开放性、同伴生产和个人赋权在建立创新型社会中发挥作用。以中国第一个黑客空间的建立为例，林特纳审视了制造商之间的合作关系并从技术使用、生产、社会、行动主义和国家之间的关系理解制造者文化[15]。赫斯塔（Herstad）使用挪威社区调查数据展示了首都企业参与创新活动的倾向性较低，但一旦参与便显现出比其他地区企业更加强烈的创新倾向[16]。从社会的视角，琼斯（Jones）等深入探索艾伯塔省德蒙顿纳米技术部门的建立过程，展示了企业家城市目标如何通过“创新”话语框架达成。这一过程涉及广泛的公民参与。此外，地方特性能帮助城市抵抗创新战略的同质化趋势[17]。在不同主体的合作方面，哈金斯（Huggins）和普罗科普（Prokop）探索了源于高校和企业构建的知识网络结构，他们发现参与者在经济发达且最具创新性的地区更有可能位于知识网络结构中具有影响力的中心位置[18]。然而，这些主体在城市更新的语境下如何作用于创新空间生产仍需期待更进一步的解读。

3. 城市更新与创新的联结

创新对技术进步的推动具有重要作用。在增长为导向的城市发展话语体系下，创新空间生产的主旨仍是提高生产效率以提升土地附加值，通过土地增值获得经济收益。由此，连接城市更新与创新空间生产的关键动力在于技术进步和土地增值：技术进步带来新的活动类型，从而对城市空间功能提出新的诉求，而土地增值则是资本逐利性推动城市空间向创新功能转型的内在动因。由此，在城市更新语境下达成创新空间生产需满足两个条件：（1）创新空间能够承载由技术进步带来的新的工作和生活模式；（2）转向创新功能后的土地产出附加值高于其原有功能类型。而厘清这一过程的发生机制则需在理解传统三方关系的基础上，即政府 - 市场 - 社会三方在更新过程中的角色参与，引入技术进步中创新主体的作用。创新主体既可以是个人，也可以是市场中具有牟利性的商业机构或具有政府背景的科研院所，因而创新主体本身也具有三方性质。基于土地政治的三方关系驱动创新空间生产，而创新主体亦对这一空间进行再塑造。

capital to promote the transformation of urban space to innovative functions. Therefore, in the context of urban renewal, innovation space production needs to meet two conditions: (1) innovation space can bear new working and living modes brought by technological progress; (2) the added value of land output after the transformation to innovative function is higher than its original function type. To clarify the mechanism of this process, we need to understand the traditional tripartite relationship, that is, the government, market and society participate in the renewal process, and introduce the role of innovation subjects in technological progress. The innovation subject can be either an individual or a profit-making commercial organization in the market or a scientific research institute with government background, so the innovation subject itself has the nature of three parties. The tripartite relationship based on land politics promotes the production of innovation space, and the innovation subject also reshaped this space.

参考文献

References

[1] Weiss, M.A. The origins and legacy of urban renewal [M]. // P. Clavel. Urban and Regional Planning in an Age of Austerity. New York: Pergamon at 56, 1988: 53-80.

[2] Harvey, D. Spaces of capital: Towards a critical geography [M]. Routledge, 2001.

[3] Robertson, K.A. Downtown redevelopment strategies in the United States: An end-of-the-century assessment [J]. Journal of the American Planning Association, 1995. 61(4): 429-437.

[4] Watt, P. 'It's not for us' Regeneration, the 2012 Olympics and the gentrification of East London [J]. City, 2013. 17(1): 99-118.

[5] Heidenreich, M., B. Plaza. Renewal through culture? The role of museums in the renewal of industrial regions in Europe [J]. European Planning Studies, 2015. 23(8): 1441-1455.

[6] Zhou, Z. Towards collaborative approach? Investigating the regeneration of urban village in Guangzhou, China [J]. Habitat International, 2014. 44: 297-305.

[7] Gainza, X. Culture-led neighbourhood transformations beyond the revitalisation/ gentrification dichotomy [J]. Urban Studies, 2017. 54(4): 953-970.

[8] Robert, P., H. Sykes. Urban Regeneration. A Handbook, British urban regeneration Association. 2000, SAGE publications. London.

[9] Davies, J.S. Partnerships and Regimes: The Politics of Urban Regeneration in the UK: The Politics of Urban Regeneration in the UK [M]. Routledge, 2017.

[10] Castells, M., P.G. Hall. Technopoles of the world : the making of twenty-first-century industrial complexes [M]. Routledge, 1994.

[11] Florida, R., P. Adler, C. Mellander. The city as innovation machine [J]. Regional Studies, 2017. 51(1): 86-96.

[12] Schumpeter, J.A. the theory of economic development [M]. MA: Harvard University Press, 1934.

[13] Huang, D., K. FAN. Gentrification or innovation oriented revitalization: discussion on emerging new urban regeneration models in China, in AAG Annul meeting 2019, American Association of Geographers: Washington, DC.

[14] Buck, N.T., A. While. Competitive urbanism and the limits to smart city innovation: The UK Future Cities initiative [J]. Urban Studies, 2017. 54(2): 501-519.

[15] Lindtner, S. Hackerspaces and the Internet of Things in China: How Makers Are Reinventing Industrial Production, Innovation, and the Self [J]. China Information, 2014. 28(2): 145-167.

[16] Herstad, S.J. Innovation strategy choices in the urban economy [J]. Urban Studies, 2017: online first, doi.org/10.1177/0042098017692941.

[17] Jones, K.E., M. Granzow, R. Shields. Urban virtues and the innovative city: An experiment in placing innovation in Edmonton, Canada [J]. Urban Studies, 2017: online first, doi.org/10.1177/0042098017719191.

[18] Huggins, R., D. Prokop. Network structure and regional innovation: a study of university-industry ties [J]. Urban Studies, 2017. 54(4): 931-952.

城中村工厂
Factory of the Village-in-city

城中村住宅
Residential of the Village-in-city

城中村住宅
Residential of the Village-in-city

城中村巷道
Roadwayof the Village-in-city

3.2

城市软更新

SOFT CITY RENEWAL

Mladen Jadric 莫拉登 · 亚德里奇

In the second half of the 19th century Vienna´s rapid growth transformed the Austrian capital into one of the largest metropoles in Europe. During this period, which lasted from 1840 until the first decade of the 20th century, also known as Gründerzeit, Vienna became one the biggest construction sites in Europe and the population grew from 400,000 to over 2.2 million citizens. Due to the limited urban space, housing became scarce and expensive despite the speed and density of construction.

In addition to this, the outdated infrastructure and particularly the poor water supply, often led to precarious hygienic conditions and the danger of epidemic diseases. In the course of the increased industrialisation and urbanisation, the construction of new districts was rushed, which certainly enabled speculative investments. A large majority of these buildings, mainly for housing, and organised in block structures, still shape Vienna's ground plan significantly. A characteristic of that particular time was the urban grid pattern of residential properties with decorated facades.

The small scale of many floor plans indicated the scarcity of residential space. Apartments normally consisted of one bedroom and a living room or kitchen. In many cases neither a lavatory nor running water was available in the apartment so that all the residents of one entire floor shared communal toilets that were situated in the corridors. Because of the horrendously high rents it was not unusual for more than six people to live in those flats as an extended family or tenants community.

By the end of the First World War the population of Vienna had declined to 1.8 million. Despite this, the housing shortage deteriorated further since the buildings had been neglected for many years during the war. Today, future developments depend on a complex web of diverse political, socio-integrative, economic and architectural measures.[1]

在19世纪下半叶，奥地利的首都维也纳经历过快速的发展，成为欧洲最大的都会区之一。从1840年到1910年，这一时期也被称为“威廉风格”时期，维也纳成为欧洲最大的建筑工地之一，人口从40万增加到超过220万。但由于城市空间的限制，尽管城市建设速度快、密度高，住房却仍旧稀缺和昂贵。

除此之外，过时的基础设施和糟糕的供水情况，经常会导致卫生条件变差和各种流行疾病的发生。在工业化和城市化进程加快的过程中，由于急于进行新区的建设，投机性的投资越来越流行。这些新建建筑大部分用于住房，并以街区结构进行组织。这些因素在很大程度上影响着维也纳的平面规划。那时期的一个特点是带装饰立面住宅形成的城市网格模式。许多建筑平面规模都很小，表明当时居住空间稀缺。公寓通常由一间卧室和一间前厅或厨房组成。公寓通常既没有厕所，也没有自来水供应。因此，整层楼的居民都共用位于走廊的公共厕所。 由于租金高得惊人，这些公寓中常有超过6个人的大家庭或房客共同居住。

第一次世界大战结束时，维也纳的人口减少到180万。尽管如此，由于这些建筑在战争期间被忽视了许多年，住房短缺的情况进一步恶化。对当今社会而言，未来的发展基于一个复合的网络，其内涵包括受复杂的政治、社会一体化、经济和建筑共同的制约。[1]

1945年之后的住宅重建政策

由于战争造成的破坏，社会住房的总体框架和居住条件发生了根本性变化。20%的住宅被破坏或损毁，特别是在工业区。另外，维也纳地区有117000所失踪房屋。因此，大量快速地进行公寓建设成为了首要目的。这导致了城市的迅速扩张：一方面体现在城市边缘地带的大型项目上，另一方面体现在城市更新和追溯致密化上。1945年至1970年间建造了超过25万套公寓，解决了住房短缺的定量目标。 在这几十年中，采取了三种方法：第一种是扩大城市规模；第二种是致密化，但要以占用旧火车站和工业设施周围市区可用的建筑

Reconstruction of residential buildings – policies after 1945

As a consequence of the destruction caused by the war, the overall framework and conditions for social housing changed fundamentally. Twenty per cent of residential houses had been damaged or destroyed, particularly in the industrial districts, and there was a shortfall of 117,000 missing homes in the greater Vienna area. The aim, therefore, was to provide as many apartments as possible and as fast as possible. This resulted in the rapid expansion of the city: On the one hand the focus was on large scale projects in the city's marginal zones, on the other hand there was also a strong focus on urban renewal and retroactive densification. With more than 250,000 apartments built between 1945 and 1970 the quantitative objectives to overcome the housing shortage were met. In the last decades three approaches were taken: the first was to expand the city, the second was densification at the expense of available building plots in urban areas around the former railway stations and industrial facilities and the third was a so called: soft city renewal. Instead of tearing down the old houses, the city of Vienna decided to refurbish the thousands of so called Gründerzeit houses. Originally built in 19- and 20 century for dense city worker settlements, through redesign and refurbishment this type of housing architecture converted urban communities into flourishing quarters for young tenants and attractive centres for creative communities.[2]

A number of them are owned by the city but many of them are privately owned. One of the examples shown in this article will explicitly explain how extensive this process is. This extends the lifespan of these already hundred year old houses for at least another fifty years.

The soft city renewal- became a very popular model that gives the streets a new dynamic and lively image. Today 800.000 Viennese are living in Gründerzeit housing blocks refurbished as part of the city´s substitution programs . Such a city structure - housing Viennese housing quartier (block) is experiencing its recovery again. It also helps the residents of Vienna to discover the advantages of a pedestrian-friendly town with short distances where life takes place in the 'Grätzeln' (Viennese dialect for city quarters). The refurbishment of former working class house settlements has become an interesting alternative to new housing constructions. It is a successful program of Viennese city urban planners in their aim to preserve the specific structure of these historic quarters based on historical preservation principles.

Soft urban renewal

From approximately 700,000 apartments at the beginning of the 1970s, around 300,000 were classified as substandard apartments, which corresponded to around 42 percent of the total stock. During this elaborate renovation of the old houses, the renovation was also usually accompanied by a change in

用地为代价；第三种就是所谓的“城市软更新”。 维也纳市没有拆除旧房子，而是决定翻新上千座“威廉风格”房屋。 这些房屋最初建于 19 世纪和 20 世纪，用于密集的城市工人住区。通过重新设计和翻新这种类型的住房，城市社区变成了以年轻租户为主体的，富有活力和创意的社区中心。[2]

其中一些房屋为市政府所有，但更多为私人所有。本文所举示例之一将体现此过程的广泛性。这将很多已有百年历史的房屋的寿命至少延长了 50 年。

软城市更新成为一种非常流行的模式，赋予了街道新的活力和生机。如今，有 800000 维也纳人居住在“威廉风格”的房屋中，这是该市替代计划的一部分。位于维也纳住宅区的城市结构正在经历再次复苏。这也有助于维也纳居民发现步行友好型小镇的优势，对以前工人阶级居住区的翻新已成为新房屋建造的有趣替代方案。这是维也纳城市规划者成功项目之一，他们的目标是在以历史保护为原则的基础上保护这些历史街区的特定结构。

城市软更新

从 20 世纪 70 年代初，在总量约为 70 万套公寓中，约 30 万套被列为不合格公寓，约占总存量的 42%。旧房精心的装修通常还伴随着前房客的改变。随着 1974 年《城市更新法》的颁布，全国城市更新指南首次在全国范围内生效。城市更新始于 1974 年维也纳第 16 区奥塔克林（Ottakring），该市决定不进行拆除旧房，而是进行一个软性的城市更新。奥塔克林作为一个试点项目，后来被当作一个模式，为许多维也纳地区的城市更新提供了动力。与居民的交流由始至终都非常重要。从一开始，维也纳的城市更新计划就蔓延到全城。1982 年《租赁法》引入了赡养费的政策，提供了新的动力。在城市软更新过程中，不仅翻修了两次世界大战期间的旧建筑，同时还包括了许多大型市政建筑，这使旧城区焕发出新的生机活力。最后，维也纳的人口发展也提出了一些必须解决具体问题：在无障碍上需要为残疾人提供便利，以及随后在老房子中安装电梯或为老年人提供新的生活形式。如今，这种支持和替代系统已经变得相当复杂，并已向各个方向扩展。 改进措施可提供有针对性的支持，例如能源（热能翻新，可再生能源的使用），生态测量（如庭院，屋顶和外墙的绿化）以及学校的安全措施（如交通通行，共用街道等）。维也纳市在这一方面堪称典范，保留并更新了其丰富的建筑遗产。但是在铁幕垮台后的 1987 年至 1993 年之间，维也纳人口在短时间内增加了大约 100000 人。这加速了城市更新的进程。数以百计的房屋、住宅区和以前的工业设施已被翻新并用于住宅。其中包括一些世界著名的项目，例如“煤气表城”项目。在欧盟的资助下，维也纳启动了最大的城市更新项目——古特尔项目——维也纳最长的基础设施带，将城市的所有火车站连接起来。古特尔区作为振兴地区工作的中心，其特点是将社会、经济、建设和文化相结合。[3]

the former tenants. With the 1974 "Urban Renewal Act", national guidelines for urban renewal came into force for the first time nationwide. The urban renewal success story finally began in 1974 with an urban regeneration of the 16th Viennese district called Ottakring. Instead of tearing down the old houses, the city decides to do a soft urban renewal. Ottakring was a pilot project and was used as a model afterwards to provide many impulses for the urban renewal of many Viennese districts. From the very beginning, contact with residents was very important. Urban renewal programs expanded all over the city when Vienna started to renovate its own buildings. The tenancy law of 1982 with the introduction of the maintenance contribution provided the new impulse. Not only the old buildings but numerous large municipal buildings from the interwar period were renovated and old town houses revitalized. Ultimately, the demographic development in Vienna also makes it necessary to address specific issues: Barrier-free living requires facilities for the disabled as well as the subsequent installation of elevators in old houses or new forms of living for senior citizens. Today, this system of support and substitution has become quite complex system and has expanded in various directions. Improvement measures provide targeted support e.g. for energy: thermal renovation, usage of renewable energy, ecological measurements like greening of courtyards, roofs and facades as well as security measures for schools such as traffic calming, shared streets and many more.

The city of Vienna exemplary preserved and renewed its rich architectural heritage. But between 1987 and 1993 after the fall of the Iron Curtain in 1989, the Viennese population increased within a short period of time by approximately 100,000 new citizens. This has given the process of urban renewal additional acceleration. Hundreds of houses, housing blocks and former industrial facilities have been refurbished and/or adopted for residential use. Among others, some of the world famous projects like The Gasometer City. With the help of EU funding, Vienna started the biggest urban renewal project - the one of the Gürtel -the longest infrastructural belt in Vienna that is linking all Railway stations of the city. The revitalization of the Gürtel-zone as a center of district work is characterized by a combination of social, economic, building and cultural measures.[3]

Within the last several decades, the focus of urban renewal has been shifted. The fields of activity have become differentiated and the range of services has become very extensive. Policy and funds for renovation became one of the central pillars of Vienna's housing policy. The housing policy makes a decisive contribution to the high quality of living with stable and affordable rents. Another important instrument of soft urban renewal was developed at the beginning of the 1990s - Block renovation. The aim was to sustainably upgrade densely built-up areas through cross-property measures - i.e. beyond the individual building and property. In 2019, for the tenth time in a row, Vienna was able to top the list of all 231 cities compared in the Mercer Study worldwide.

在过去的几十年中，城市更新的重点已经转移。活动领域已经变得多元化，服务范围也变得非常广泛。改造政策和资金成为维也纳住房政策的中心支柱之一。为确保居民能获得稳定且负担得起的高质量生活住房，政府住房租金政策起了决定性的作用。20 世纪 90 年代初，政府推进了另一项重要的城市软更新手段：街区改造。目的是通过跨财产措施（即超出单个建筑物和财产），可持续地升级密集的建筑区域。在 Mercer 研究的全球比较中，维也纳在 2019 年连续第十次荣登全部 231 个城市的榜首。维也纳的“城市软更新”已无数次获得国际奖项。2010 年，维也纳市在联合国开发和更新的城市发展和更新领域获得了联合国人居署荣誉奖。城市更新反映了维也纳市的杰出力量：对当地特色的支持，对历史建筑的尊重，以及与当代建筑和建筑标准的独特结合。[4]

改造前的“灰姑娘”
The "Cinderella" before reconstruction

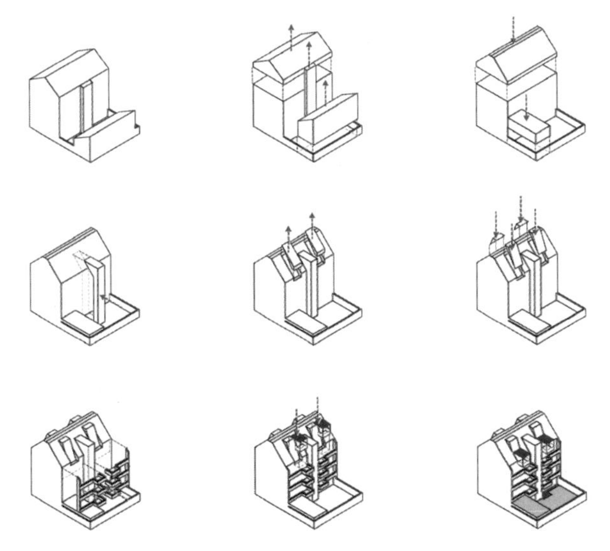

改造步骤：拆除、阁楼转换、电梯、院子扩建、梯田、绿化
Renovation steps: demolition, loft conversion, elevator, yard extension, terrace, landscaping

Vienna's Soft Urban Renewal has been the recipient of international awards for the umpteenth time. In the year 2010, the city of Vienna was awarded with. the UN-Habitat Scroll of Honor Award in the field of urban development and renewal by the UN World Organization for Settlement and Housing. Urban renewal mirrors the outstanding strength of the city of Vienna: the support for local identity, respect to historical building stock and unique combination with contemporary architecture and architectural standards.[5]

A case study - Cinderella from Ottakring

Houses like this one are characteristic for several Viennese districts. The process of renovation can be illustrated using the example of this case study. Typical Gründerzeit-houses are 4 to 5 floors high and, as a rule, built from bricks. The roof structure and floor slabs are made out of wood with a roof covered with tiles. Decor of the facade could no longer be renewed after the Second World War and over time the facade was knocked off and simply plastered. A particular house in the former industrial cluster in Ottakring was in totally desolate condition: the structure became unstable and the infrastructure of the house was completely ruined.

The design concept of the outer shell includes subtle changes to the street but complete changes of the outer skin to the green courtyard. With an almost therapeutic-color concept and yellow (the sun-colored) entrance and staircase, the warm and optimistic atmosphere spread insideout. The inner courtyard remains the green courtyard, as a private garden for the tenants. The whole house has an intimate scale and combines the qualities of the apartment block with the scale of a single-family house. The neighborhood of the house was also completely renewed as a result of the upgrading in recent years around the Ottakringer brewery and chocolate factory Manner and as numerous mobile young group of tenants started looking for an interesting and inexpensive place to live. When we started with a renovation of the house at Gansterergasse street Nr.14 we found a typical former worker´s house in the old industrial district in a very desolate condition and the yard clogged with all sorts of garbage on the ground floor. The renovation, and partly reshaping, of the house was shown in nine steps. The house was changed and expanded by adding a Penthouse on the roof - as well as through a one-story extension in the courtyard, new elevator and balconies. Particular attention was paid to improving the quality of the residential spaces – and as the diagrams show- it was a rather subtractive strategy based on the selected principles listed later on. The quality of outdoor spaces: courtyard, large loggias, roof garden on the extension, big terraces on the roof, determine the design. Despite the small, almost intimate scale of the project, a high spatial experience was created. The apartments in the street wing (ground floor) are studios for homeworkers with an extension to the courtyard. A spacious garden – an internal courtyard- unites the complex. All apartments are enjoying benefit from the generous

案例研究 奥塔克林的“灰姑娘”

改造后的灰姑娘
The "Cinderella" After renovation

像这样的房屋是维也纳几个地区的特色。可以使用此案例研究的示例来说明翻新过程。典型的“威廉风格”房屋高 4、5 层，通常由砖砌而成。屋顶结构和楼板由木材制成，屋顶覆盖着瓷砖。在第二次世界大战后，立面的装饰不能再更新，随着时间的推移，立面被拆除并简单地抹上灰泥。奥塔克林（Ottakring）前工业区中的一所房屋完全处于荒凉状态：结构变得不稳定，房屋的基础设施遭到完全破坏。

立面改造前
before the renovation of facades

建筑外形的设计理念考虑了对街道的细微改动，但对绿色庭院的外形进行了整体改动。几乎全部采用了治疗性的颜色概念，黄色（阳光色）的入口和楼梯，温暖而乐观的氛围由内而外散发。内院仍然是绿色庭院，是租户的私人花园。整个房子的建设同时考虑了公寓的质量和单户住宅的规模，整体设计显得亲密无间。由于近年来奥塔克林啤酒厂和巧克力工厂进行了升级，住宅周围的社区也得到了彻底的更新，许多年轻的租户开始寻找一个有趣的、便宜的地方居住。当我们开始对 Gansterergasse street 14 号的房子进行翻修时，我们发现在老工业区有一个典型的前工人居住的房子，非常荒凉，一楼的院子里堆满了各种垃圾。这座房子的翻修和部分翻修分 9 步进行。通过在屋顶上增加一个阁楼，对院子中的一层楼扩建，增设新的电梯和阳台，完成了房屋的改造和扩建。特别关注于提高居住空间的质量——正如图表所示——这是一个相当大的减法策略，基于后面列出的选定原则。室外空间的质量：庭院、大型凉廊、延伸的屋顶花园、屋顶的大露台等因素决定了设计方向。尽管项目规模很小，但却创造了舒适的空间体验。临街侧（一层）的公寓是家庭工作者的工作室，延伸至庭院。一个宽敞的花园——一个内部庭院，两者联合了这个综合体。所有公寓都可从宽敞的新落地玻璃窗中受益，这些新的落地窗可通往街道，并且在后侧显得格外宽敞。 每间公寓都经过仔细考虑。通过合并每个楼层平面图上的小公寓，公寓变得更大了，并获得了所有必要的舒适度。在建筑师、建筑商和市政当局 (WohnFond Wien) 的密切合作下，每一个细节都得到了开发和优化。通过额外的元素和扩展，每一套公寓都有一个室外空间或花园，上层有大的凉廊，而在一楼的公寓则设有私密的露台或花园。在屋顶上，我们创建了两个宽敞的顶层公寓 (Maisonette's)，露台面向大的绿色庭院。

立面改造中
During the renovation of facades

立面改造后
After the renovationof facades

曾经一栋荒凉的房子，满是垃圾的肮脏的院子，现在蜕变成一个生活的天堂。在童话中，我们称其为奥塔克林的“灰姑娘”。

new French windows to the street and on the back side exceptionally spacious loggias. Each single apartment is considered carefully. Through the merging of small apartments on every floor plans, the apartments are much bigger and got all necessary comfort. In close cooperation between architects, builders and city authorities (WohnFond Wien) every single detail has been developed and optimized. Through additional elements and extensions, every single flat has an outdoor space or garden with big loggias on upper floors and intimate and quite patios/ gardens for apartments on the ground floor. On the roof we created two spacious penthouse flats (Maisonette´s) with terraces facing the big green courtyard of the block.

A house in a desolate condition with a formerly dirty courtyard and used as a rubbish yard became a small paradise for living that we, alluding to the fairytale, called Cinderella from Ottakring.

On-site improvement work:

• Extension of a passenger elevator with 5 stations
• Central heating with gas condensing technology
• Central water heating system
• manufacture of burglar-resistant apartment entrance doors (WK 3)
• Execution of the street-side windows as soundproof windows-at least RW = 38dB
• Set up a central satellite system including piping and cabling for all apartments
• Manufacture of empty piping for cable TV and telephone
• Manufacture of a lightning protection system
• Production of a gate intercom
• Installation of a garbage room
• Design of the inner courtyard
• Manufacture of additional riser and distribution lines for electricity and water
• Reinforcing or redesigning waste strands
• Greening the flat roof of the courtyard extension

Conservation measures:

• Repair all facades
• Thermal measures:
• Thermal insulation of the courtyard facades
• Thermal insulation of the street facade
• Installation of heat protection windows
• Thermal insulation of the roof
• Thermal insulation of the ground floor floor against the basement
• Restoring basement ventilation
• Reconstruction of the house entrance and yard gate
• Masonry drainage
• Underpinning of foundations / new floor slab on the ground floor made of reinforced concrete

现场改进工作：

• 扩建 5 站式的客梯；
• 采用气体冷凝技术的集中供暖；
• 中央热水系统；
• 制造防盗公寓入口门；
• 街道侧窗采用隔声窗，至少计权隔声量 RW=38dB；
• 建立中央卫星系统，包括所有公寓的管道和电缆；
• 制造有线电视和电话的空管道；
• 制造防雷系统；
• 生产门禁对讲机；
• 安装垃圾房；
• 内部庭院设计；
• 制造用于水和电的附加立管和配电管线；
• 加强或重新设计废物链；
• 庭院内屋顶绿化。

保护措施：

• 修复所有立面散热措施；
• 庭院外墙的隔热；
• 街道外墙的隔热；
• 安装隔热窗；
• 屋顶隔热；
• 底层对地下室的隔热；
• 恢复地下室通风；
• 房屋入口和院门的重建；
• 砌体排水；
• 钢筋混凝土底层的地基 / 新楼板的支撑；
• 根据静力学要求更换顶棚；
• 恢复屋顶覆盖层；
• 恢复屋顶和立面面板；
• 制造新的双通风收集烟囱；
• 根据区域供暖的需要而设计的中央供暖壁炉；
• 楼梯和走廊的维修；
• 修复公寓入口门和普通房间入口门；
• 修理地下室；
• 在庭院中创建新的存储空间；
• 新蒙太奇的宫廷石膏和绿地；
• 围栏（围墙）、新花园围栏的维护；
• 更新立管和配电线路；
• 为中央供暖系统建立新的燃气连接；
• 用新的塑料立管和分配管替换所有水管——不再使用铅水管；
• 废水管的更新；
• 修理房屋管道连接；
• 所有墙壁的正常维修。

- Ceiling replacements according to statics
- Restoring the roof covering
- Restoring the roof and facade panels
- Manufacture of new double-draft collecting chimneys
- A central heating fireplace according to the needs of the District heating (Fernwärme)
- Repair of staircase and hallways
- Restoring the apartment entrance doors and the entrance doors to the general rooms of the house
- Repair the basement
- New creation of storage spaces in the yard
- New montage of the court plasters and the green areas
- maintenance of the fencing (boundary walls), new garden fence
- Renewing the riser and distribution lines for electricity
- Establishing the new gas connection for the central heating system
- All water pipes against new riser and distribution pipes made of plastic replaced - no more lead water pipes available
- Renewal of the waste pipes
- Repair the house duct connection
- General repair of all walls

参考文献

References

[1][2] 2019 M.Jadric, D.Alic (Eds.).At home in Vienna : Studies of Exemplary Affordable Housing, Vienna : TU Wien Academic Press Wien, 2019.

[3] GB* ZUSAMMEN BESER LEBEN, Stadt Wien, Technische Stadterneuerung, November 2019.

[4] GB* Gebietsbetreuung Stadterneuerung, ABSCHLUSSBERICHT 2012 - 2017, MA 25 - Stadterneuerung und Prüfstelle für Wohnhäuser, Maria-Restituta-Platz 1, 1200 Wien

[5] 2018 M. Jadric: .Asien: Wohnen in China, Indien, Indonesien, Singapur und Südkorea. in: “Das Wiener Modell 2”, W. Förster, W. Menking (Hrg.); Jovis, Berlin, 2018: S. 28 - 43.

公寓照片：改造前、改造中、改造后

Apartment photos - before, during and after the renovation

屋顶照片：改造前、改造中、改造后

Photo of the roof - before, during and after the renovation

麻磡村位于国家级高科技产业走廊上，毗邻城市中心，却没有得到很好的发展。地缘优势给予麻磡未来发展的无限可能，改造实验以此为题，研究团队实地调研、重新定位、整体规划，提出改造策略、运营策划、实施路径等多个维度的设计研究。

Makan village is loacted in the high-tech corridor of China while neighboring the city centre. It is not well developed despite its geographic advantages. The research is aiming to achieve the potential of the village, conducting on-site investigation, repositioning the renewal and making an overall planning. A design, studying the process of renewal methods, operation planning and implementing approach, is put forward.

4 麻磡村的改造实验——愿景

城中村综合整治的整体规划

MAKAN RENOVATION STUDY—VISION

THE OVERALL PLANNING OF URBAN VILLAGES RENOVATION

4.1 麻磡村地理位置
LOCATION OF MAKAN VILLAGE

麻磡村位于深圳市南山区北部，隶属于水源三村。水源三村是靠近西丽水库水源保护区域的三个村落群，麻磡村位于水源三村的中部，它的西部是白芒村，东部是大磡村，北部是羊台山公园，南部是西丽水库。本次研究范围是麻磡村全域，如图中白色区域所示，用地面积为7.1km^2。

Makan Village is located in the Nanshan District, north of Shenzhen city center. It belongs to the Shuiyuan Sancun, a group of three villages in the water, air and nature protection area close to Xili reservoir (Xili Lake). Makan Village is located in the middle of the Shuiyuan Sancun, between Baimang Village in the west and Dakan Village in the east. North of the planning area is the Yangtai Mountain and in the south the Xili reservoir. The planning area is the whole Makan Village, seen in white, which is 7.1 km².

区位图
Location

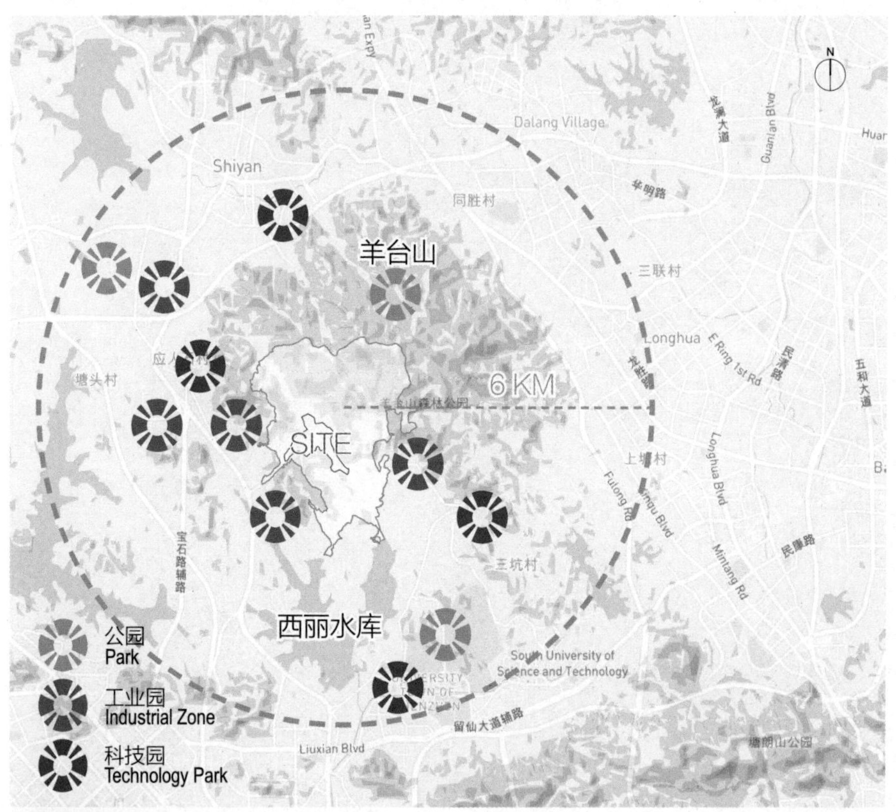

基地周围环境
Surroundings

4.2 历史,机遇与挑战
HISTORY, OPPORTUNITIES AND CHALLENGES

麻磡村曾名“麻冚围”——《新安县志》客家后裔迁徙于此，建民居，有“麻石半墙，青砖到栋”的风貌。

Makan village was once named Makanwei. According to the records, its name originates from the granite walls and black bricks in the village.

1819

回村的南洋华侨拿出多年积蓄，组织村民修建碉楼，用来打击进村抢劫的土匪。

The overseas Chinese in Southeast Asia donated their savings and organized villagers to build watchtowers to fight against bandits.

1926

“第四战区游击纵队指挥所第四游击挺进纵队直辖二大队”在麻磡，进行艰苦抗日斗争。

People are fighting against the Japanese during the WW Ⅱ.

1959

“二线”交付使用，“二线”穿过麻磡村，既是特区管理线，又是边境管理线。

The village lied on the management line of special zone and border management line.

1986

“深圳大学城”开始建设，次年，教学基础设施建设并入驻。

The “Shenzhen University Town” started to be built. In the following year, the construction of educational infrastructure was basically completed.

2002

麻磡村历史沿革示意图

The timeline of Makan Village

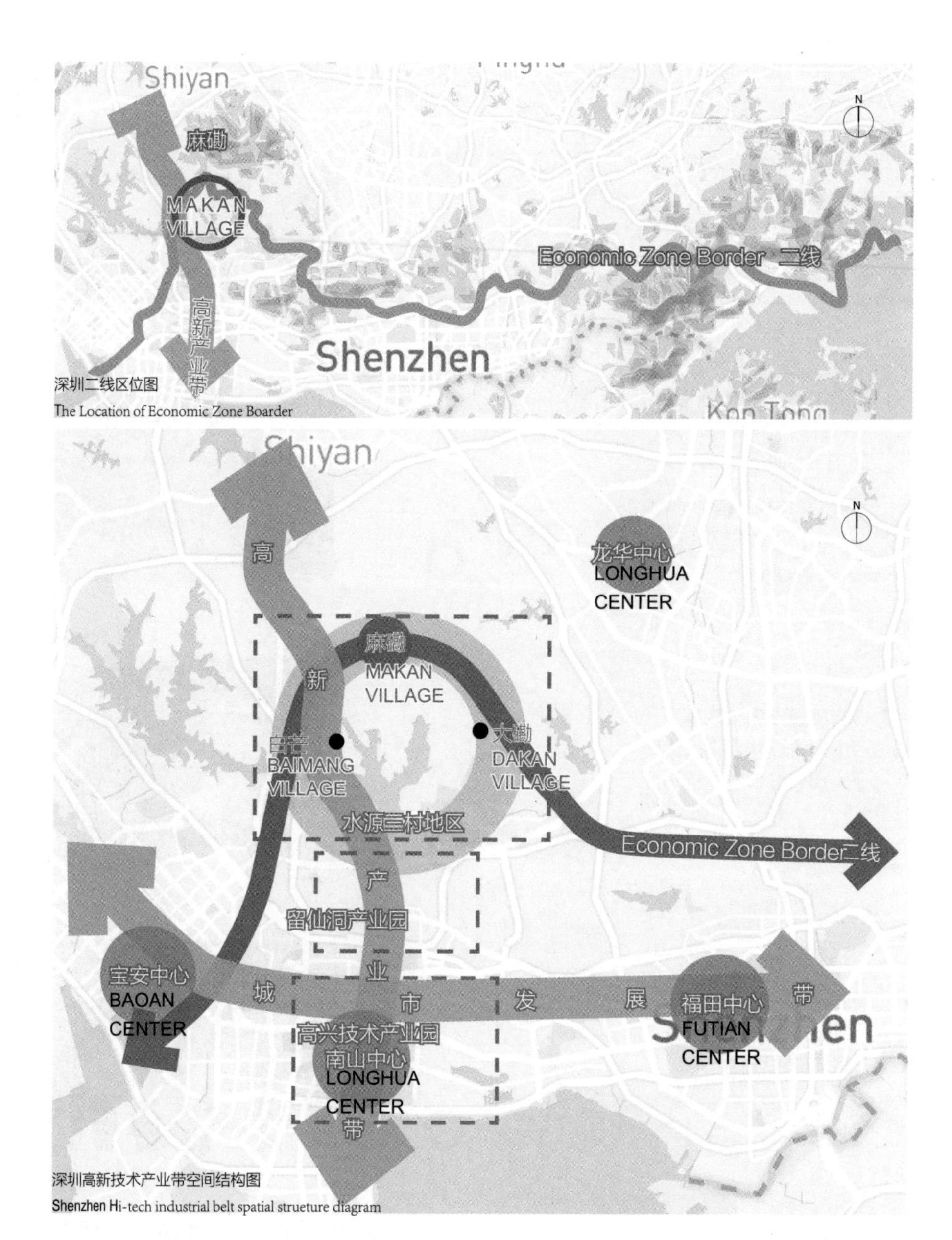

深圳二线区位图
The Location of Economic Zone Boarder

深圳高新技术产业带空间结构图
Shenzhen Hi-tech industrial belt spatial strueture diagram

水源三村位于深圳市城市发展时间和空间的一个横断面——二线，以山湖村落为特色，是弥合关内外分隔的重要节点之一。同时，水源三村位于二线与高新产业带交汇点上，是水源三村发展的区位基础。

Shuiyuan Sancun is located in a cross section of Shenzhen's urban development time and space—Economic Zone Border,with the characteristics of mountain and lake villages,which will become one of the important nodes to bridge the inner and outer separation.

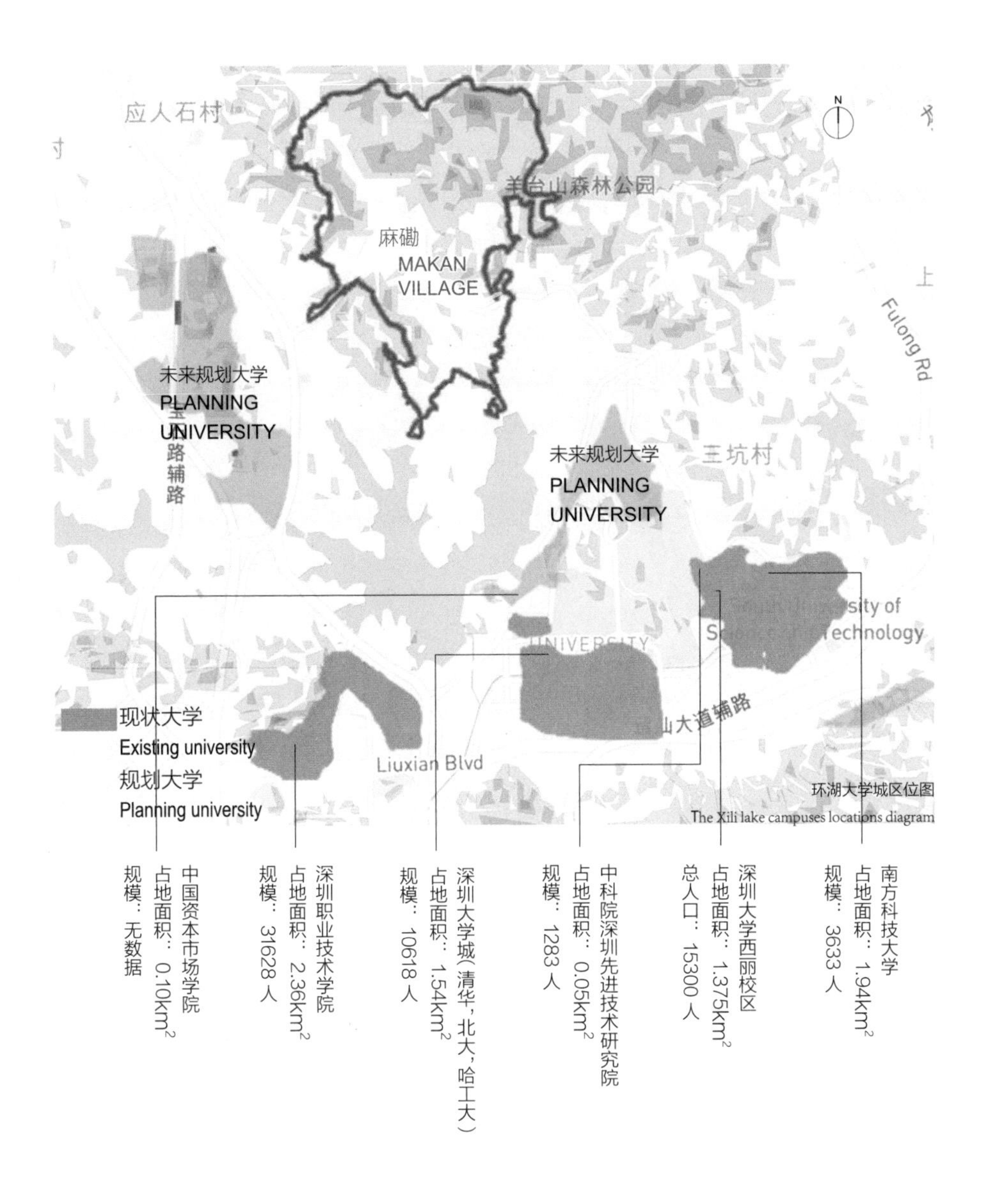

环湖大学城区位图

The Xili lake campuses locations diagram

现状大学与规划大学环西丽湖成片分布，形成环西丽湖大学城。现状大学主要分布在西丽湖南侧，师生规模约6万人。规划大学主要分布在西丽湖东西两侧，因此项目基地周边聚集了丰富的高校资源。

The current university and the planned university are distributed, forming a ring of XIli Lake Campuses. The current status of the universites are mainly distributed in the south side of Xili Lake, with staffs and students counting to about 60,000. The planned universities are mainly distributed in the east and west sides of Xili Lake, which has enriched university resources around the Makan.

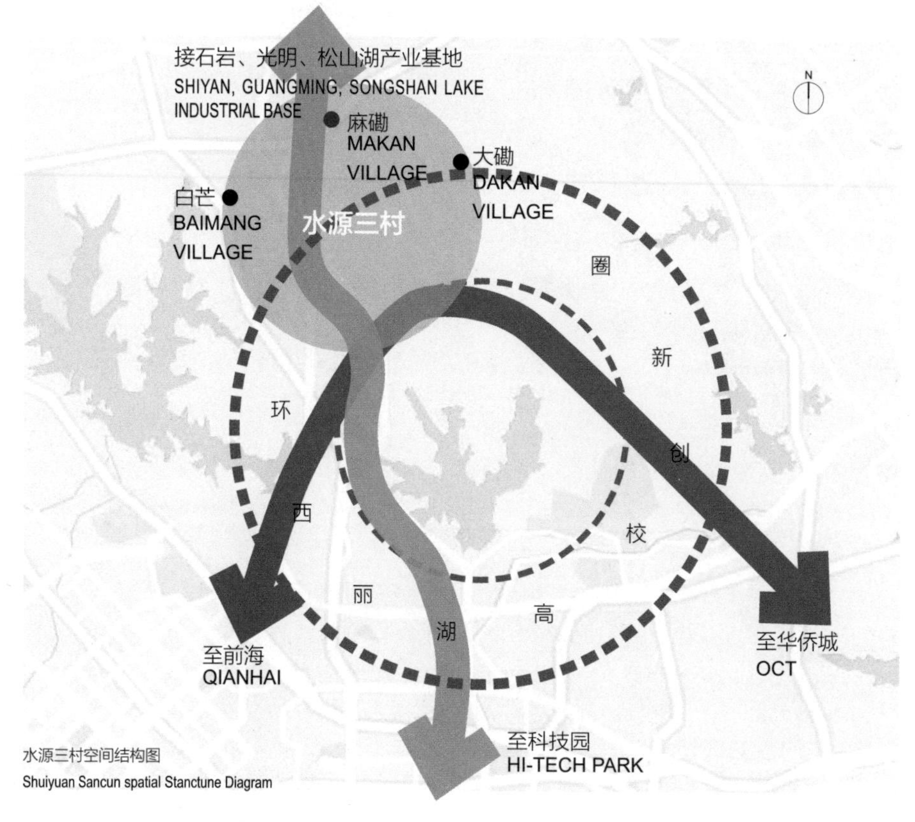

水源三村空间结构图
Shuiyuan Sancun spatial Stanctune Diagram

现状多所大学与未来规划大学环西丽水库建设，形成环西丽湖大学圈。水源三村将与周边大学构成环西丽湖高校创新圈。

At present, many universities and future planned universities are built surrounding the Xili Reservoir, forming a circle around the Xili Lake Campuses. Shuiyuan Sancun will form a circle of innovation with the surrounding universities.

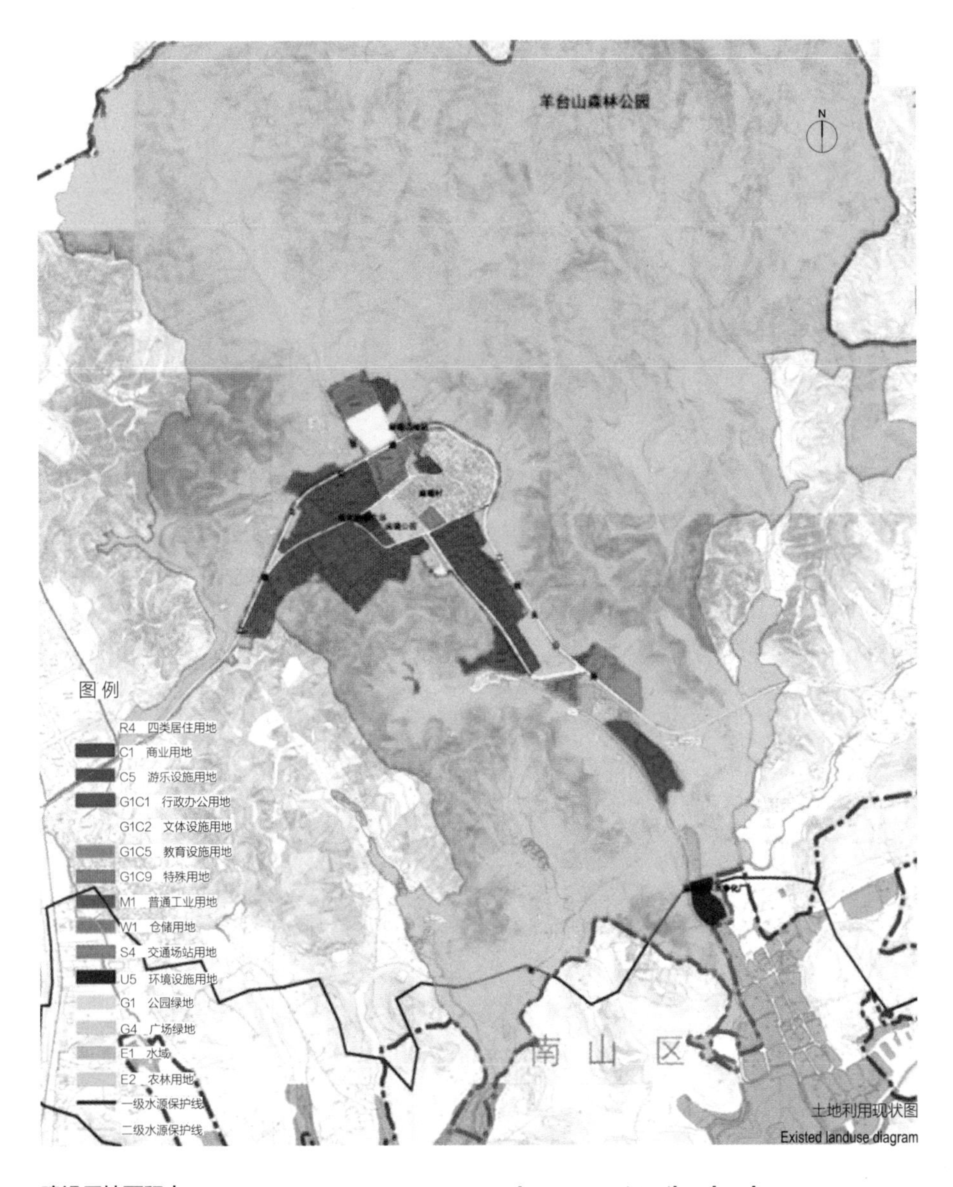

土地利用现状图
Existed landuse diagram

建设用地面积少

麻磡村全域面积为 714.09hm²。建设用地面积为 55.94hm²，占总用地面积的 7.8%。非建设用地面积为 658.15hm²，占总用地面的 92.2%。 建设用地主要为工业用地，用地面积为 25.99hm²，占建设用地面积的 46.5%。

Less construction land area

The total area of Makan Village is 714.09 hm². The construction land area is 55.94 hm², accounting for 7.8% of the total land area. The area of non-construction land is 658.15 hm², accounting for 92.2% of the total land. Industrial land is the main construction land-use, with 25.99 hm², accounting for 46.5% of the construction land area.

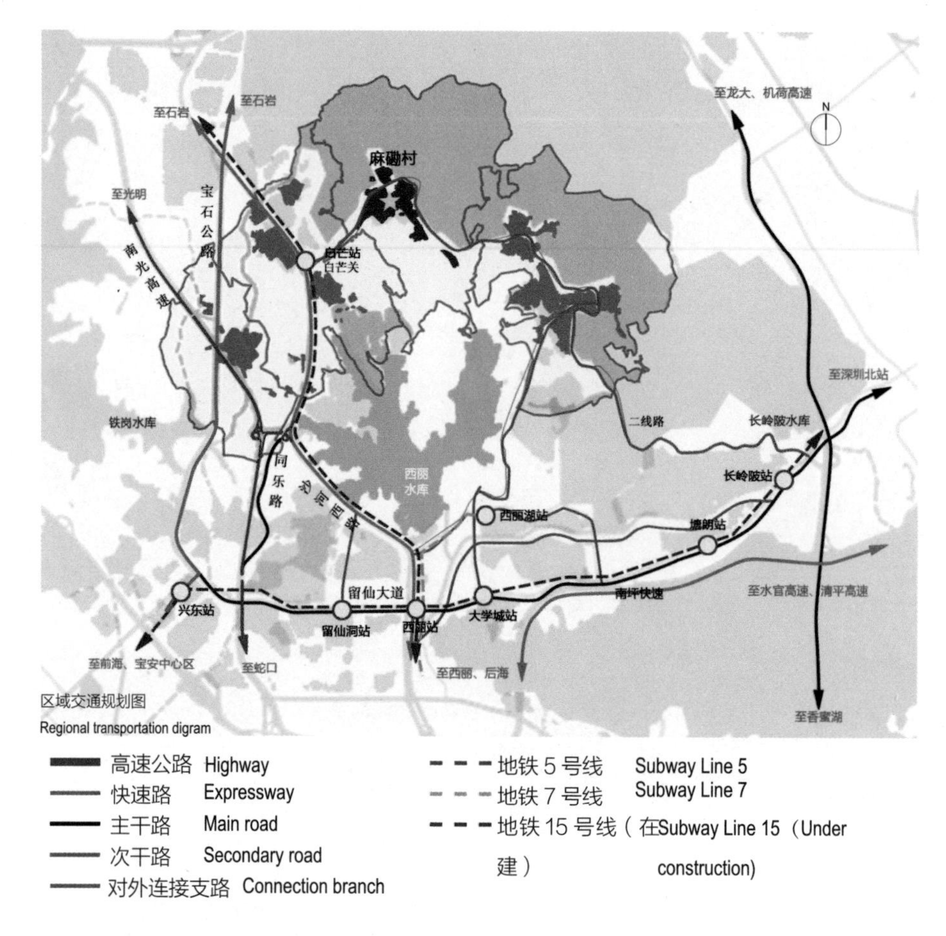

区域交通规划图
Regional transportation digram

两条线路连接外部主要城市道路，未来会有新的地铁站点。

Two subway lines connect external main city roads There will be new subway stations in the future.

基本生态控制线
Basic ecological control line

一级水源保护区
Primary water source protection area

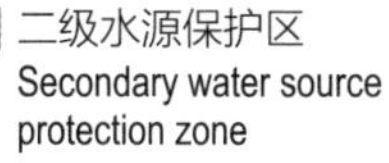

二级水源保护区
Secondary water source protection zone

环境空气质量功能区
石岩水库—铁岗水库及西丽水库一类区
Shiyan reservoir - Tiegang reservoir and Xili reservoir Air quality functional area

区域 Area	用地面积 Land area	比例 Proportion
一级水源保护区 Primary water source protection zone	18.93hm^2	2.65%
二级水源保护区 Secondary water source protection zone	694.34hm^2	97.23%
水源保护区外 Outside the water source protection area zone	0.82hm^2	0.12%
合计 Total	714.09hm^2	100%

基本生态控制线

麻磡村全域范围均在基本生态控制线之内。境内设施对生态环境、水源保护和河道管理、森林资源保护无不利影响，允许按现状、现用途保留使用。

空气质量功能区

麻磡村全域用地均在一类环境空气质量功能区（石岩水库—铁岗水库及西丽水库一类区）范围内。需执行国家《环境空气质量标准（GB　3095- 设施 012）》规定的一级标准。

饮用水源保护地

麻磡村 97% 的用地处于二级水源保护区内，3% 的用地位于一级水源保护区。

Basic ecological control line

The entire area of Makan Village is within the basic ecological control line. Facilities will be preserved without adverse impact on the ecological environment, water source protection and river management, and forest resource protection.

Air quality functional area

The whole land in Makan Village is within the scope of a class of ambient air quality functional zones (Shiyan Reservoir-Tiegang Reservoir and Xili Reservoir).

Drinking water source protection

97% of the land in Makan Village is in the secondary water source protection zone, and 3% of the land is located in the primary water source protection zone.

4.3 现状分析
SITUATION ANALYSIS

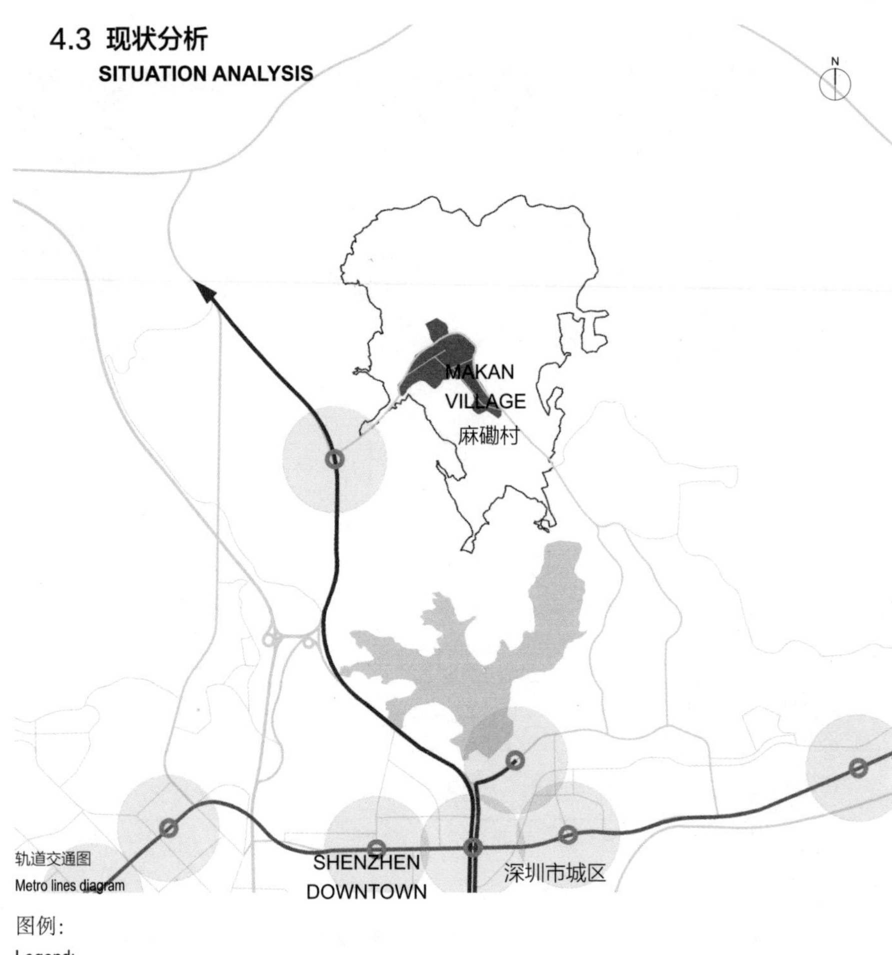

轨道交通图
Metro lines diagram

图例:
Legend:

地铁 5 号线	Line 5	地铁站	Subway stop
地铁 7 号线	Line 7	800m 辐射圈	800m radius
地 铁 13 号 线（规划）	Line 13 (planned)		

周边地铁网

深圳作为地铁系统极为发达的城市，正在向北扩展和增密其地下交通系统，这促使了目前计划的 13 号线的发展，该线路在主要道路附近会设有一个地铁口， 从地铁口到麻磡村西部步行约 20 分钟。目前有一条公交线路，缩短了地铁口到麻磡的时间。

Surrounding subway network

As a city with highly developed subway system, Shenzhen is expanding and densing its underground transportation system to the north, which has prompted the development of the currently planned Line 13. The line will have a subway exit near the main road and a 20-minute walk from the subway to the west of the village. However, due to the current bus route, this shortens the travelling period from the subway to the Makan Village.

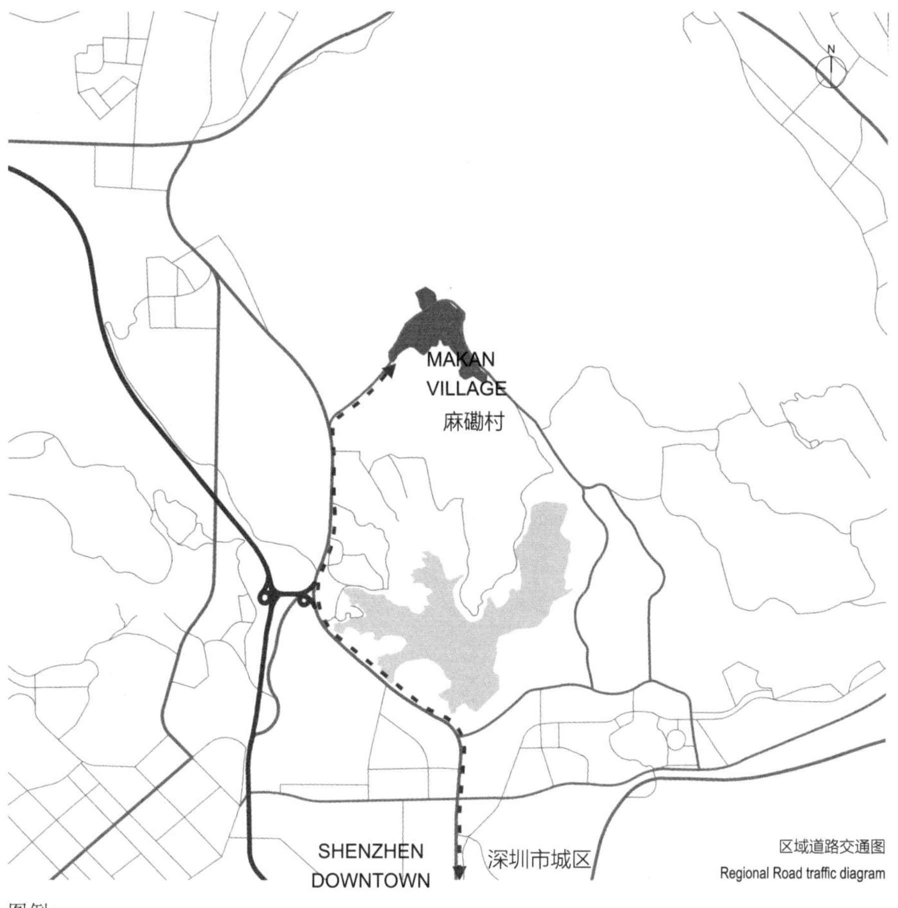

区域道路交通图
Regional Road traffic diagram

图例：
Legend:

高速路	Highway	城市支路	Branch road
快速路	Freeway	最短路线	
城市主路	Main road	ShortWest route between Makan Village & Shenzhen Downtown	

周边道路网

麻磡村通过一条主要公路与外界相连，该道路沿着村庄北侧环绕，不会扰乱村庄内的交通流线。这条公路也将麻磡村与周围两个相邻的村庄联系起来，但更重要的是这条道路连接了 S33 南光高速公路和通往深圳市中心的道路。西边的沙河西路是去深圳市中心最快的路，距市中心 40 分钟车程。

Overall Road Network

Makan Village is connected to the surroundings by a main road which goes around the village on the north side, not disturbing the traffic inside the village. This road connects Makan Village to the two neighbouring villages,but more importantly to the S33 Nanguang Expressway and to roads leading to Shenzhen downtown. The fastest connection to Shenzhen downtown is from the west along the Shahe West Road and takes about 40 minutes.

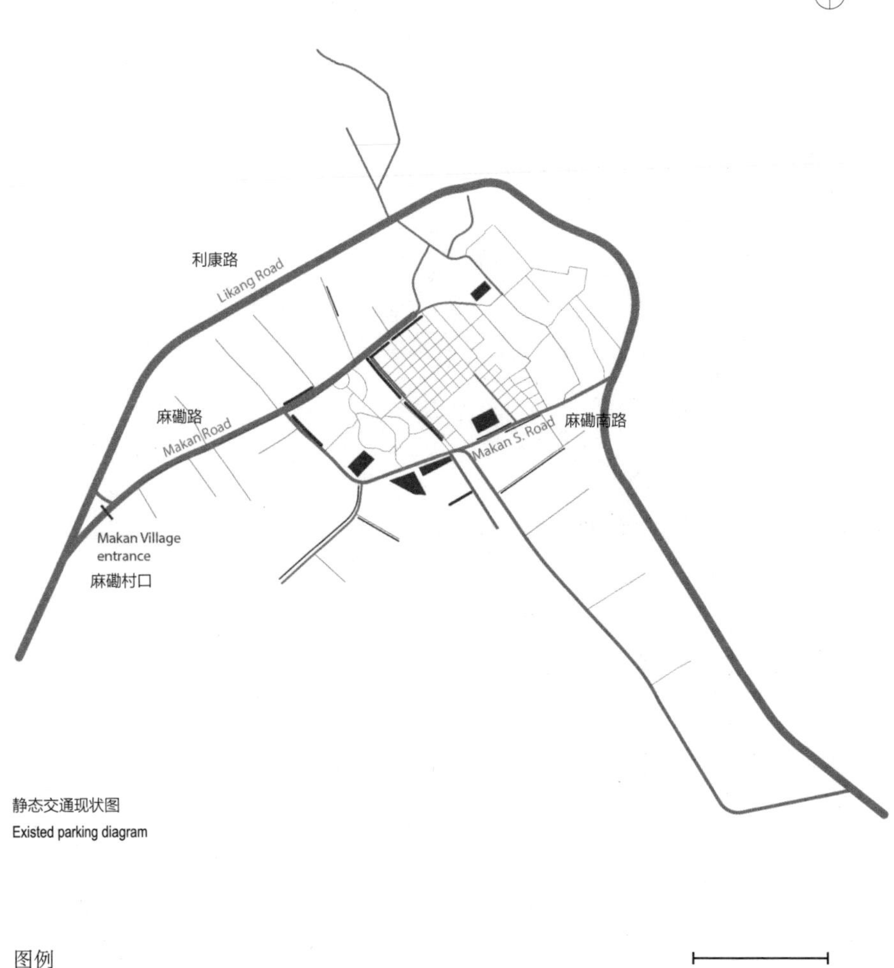

静态交通现状图
Existed parking diagram

图例
Legend:

- 道路 Street
- 停车 Parking

道路系统

麻磡村位于区域的北侧，东部被利康路围绕。利康路连接村庄和深圳市中心。该村的道路网络基于两条主要道路和几条连接住宅区和工厂的小型道路。主要道路也是该村非常重要的商业区。如上图所示，村里的大部分停车位都位于这些主要道路上。

Road System

The village itself is in the north and the east enclosed by Likang Road. It connects the village with the downtown of Shenzhen. The road network in the village itself is based on two main roads and several small roads to connect residential areas and factories. The main roads are also a very important commercial area for the village. As the map above shows, most of the parking space in the village is located at these main roads.

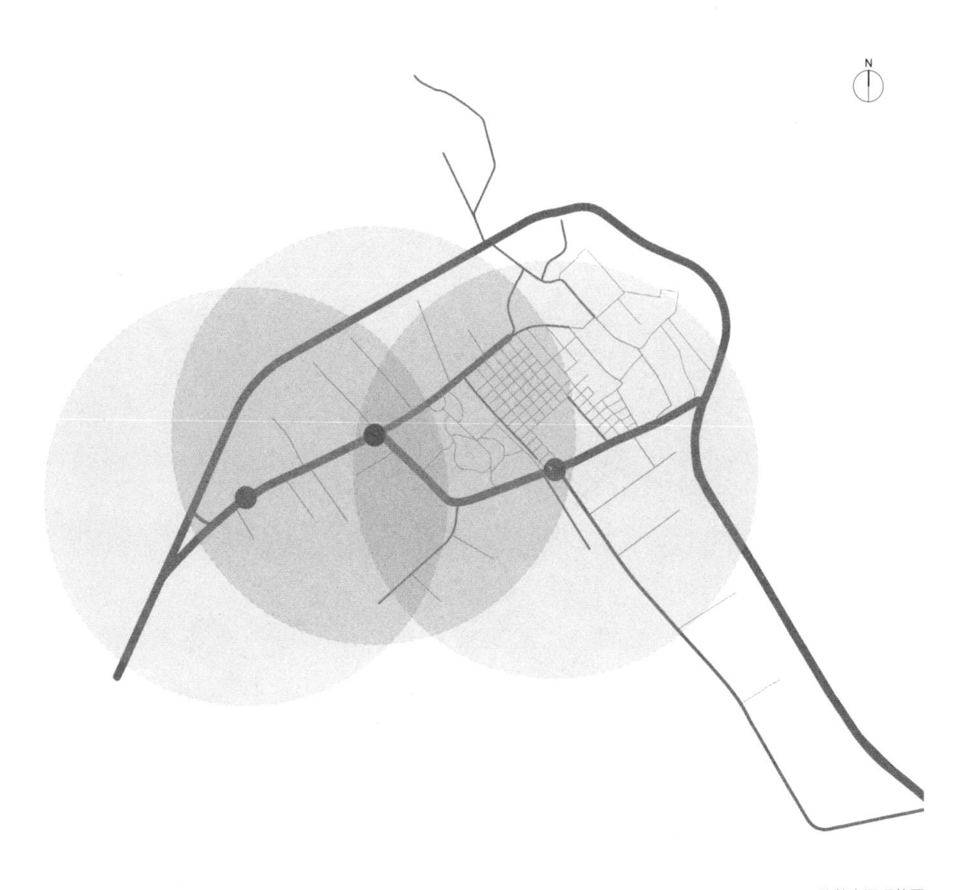

公共交通现状图
Existed public transport diagram

图例
Legend:

主要道路 Main Road
300m 辐射圈 300m Radius

0 200m

公共交通

公共汽车覆盖了麻磡村的大部分地区，因此是该村最重要的资源之一。很多地区都有多个公交车站，通过较小的调整，可以更有效地利用公共交通系统。由于公交系统连接了城中村与市中心，同时更为了让居民更方便地到达新地铁口，因此这一调整至关重要。

Public Transport

The bus covers most of Makan Village and is therefore one of the most important assets of the village. However a lot of the area is served by multiple bus stations. With small adaptions the public transport system can be managed more efficiently. This is critical since the public transportation is an important connection to Shenzhen's downtown and will be even more important with the new subway station.

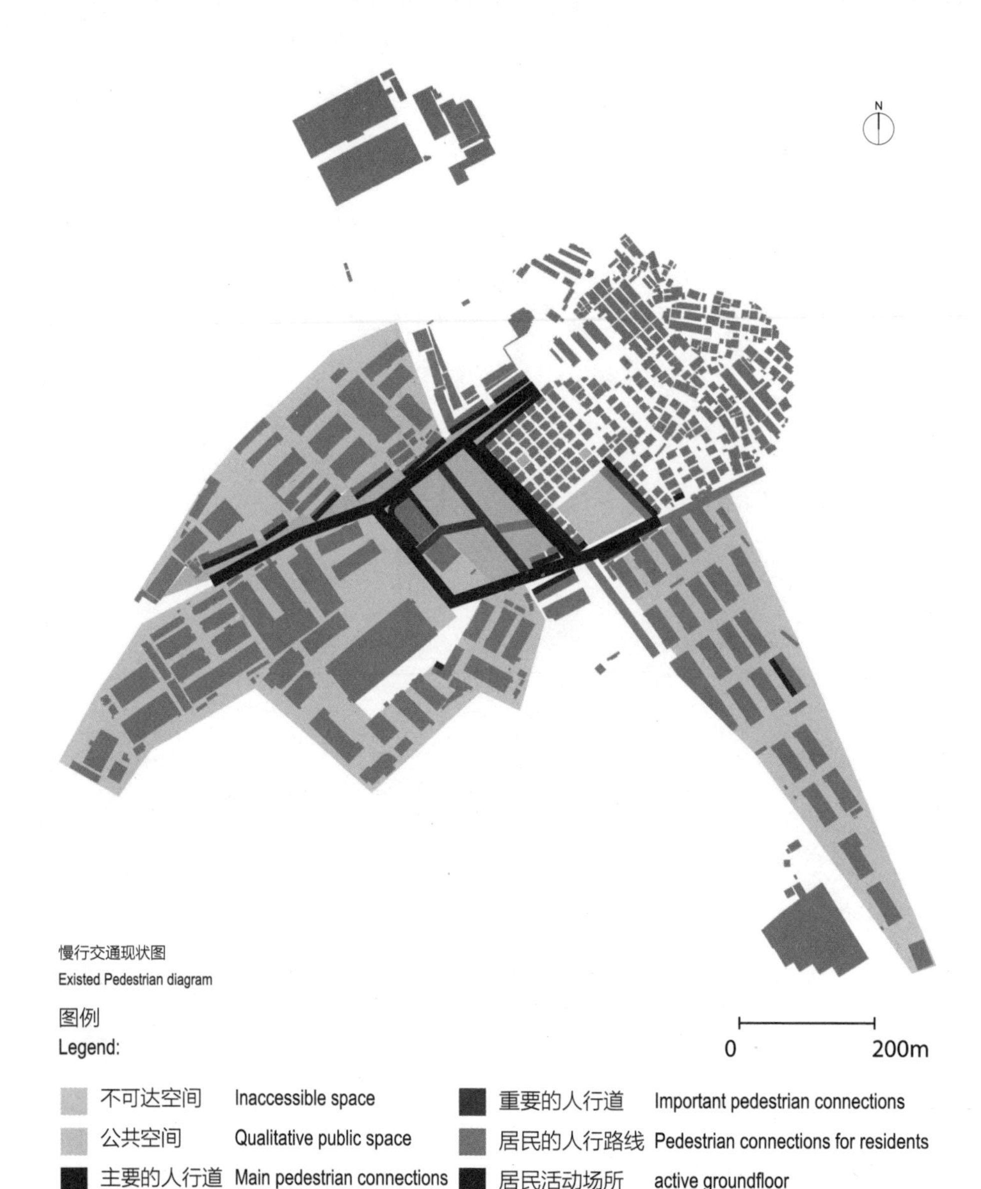

慢行交通现状图
Existed Pedestrian diagram

步行与活动空间

目前，该村的特点是公园周围有一个公共核心，像商店这样的地面活动集中在这个核心。除了东北部的住宅区外，公共核心区域围绕栅栏设置。由于周边地区的许多场所均设门禁系统包括工厂等，村中区域过于碎片化，各区域可达性差，阻碍了城市活力空间的形成。

Pedestrians & Active Groundfloor

Currently the village is characterized by a public core around a public park. The groundfloor activities like shops are concentrated in this core. The areas around the public core are gated behind fences, with the exception of a residential area in the northeast. A lot of the gated areas include factories. The fragmented accessibility pattern puts restrictions on the production of a dynamic urban life.

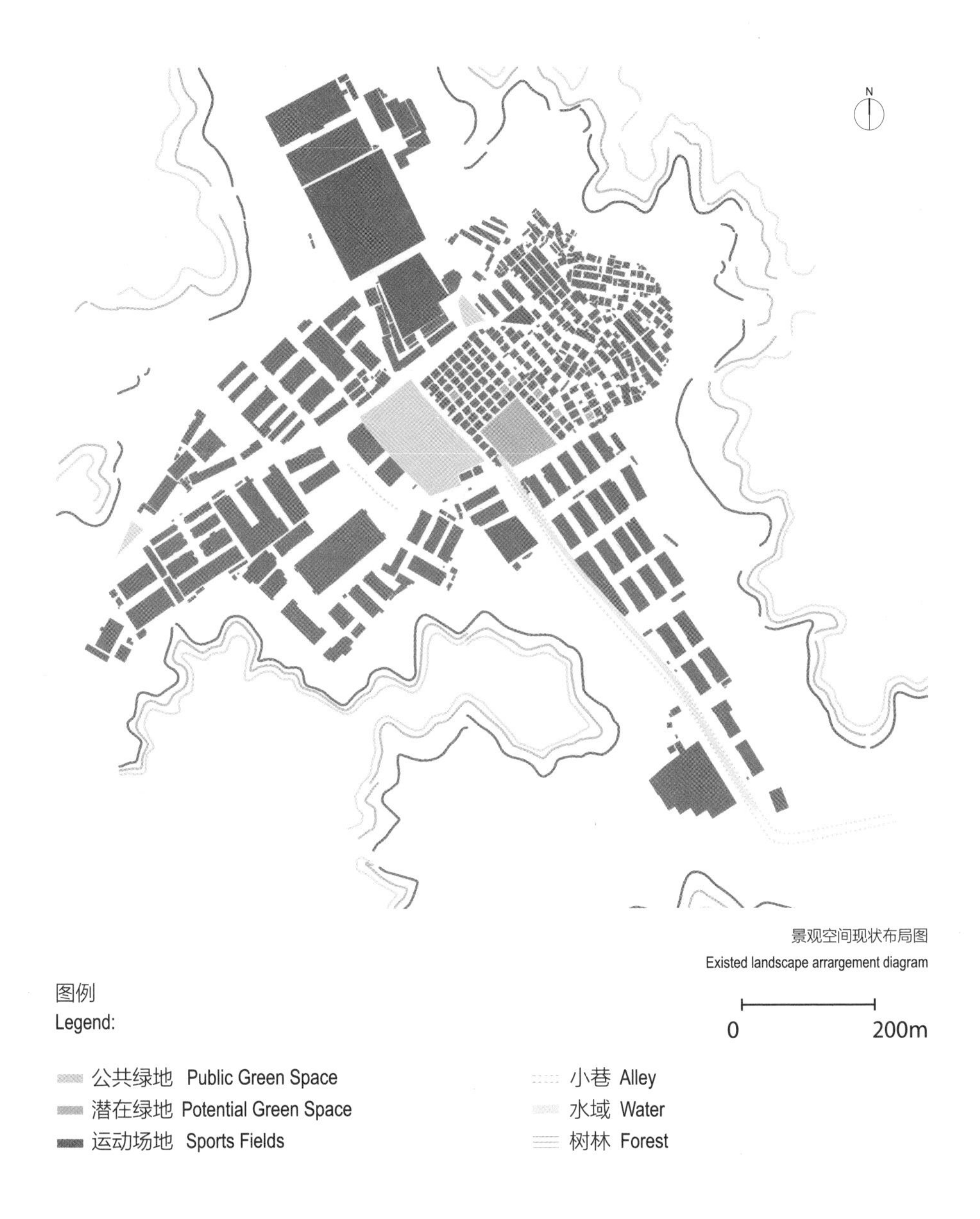

景观空间现状布局图
Existed landscape arrangement diagram

绿化与水景空间

村内只有一小部分的绿地可供公众使用，运动场的空间有限，村中心的空旷绿地目前被用作临时停车场和垃圾收集站，无法发挥其潜力。 在这里被发现的潜力空间是位于村庄西北侧被居民用来种菜的荒地。另一个潜力要素是一条小河，主要位于地表之下，由北向往南流入树林。

Green and Blue Space

Only a small part of the village's green space is accessible to the public. The access to sports fields is limited and the big empty green space in the center of the village does not reach its potential, as it is currently used as an informal parking lot, trash collection site. The potential we see here is the north-west side, which is used for gardening by the residents.

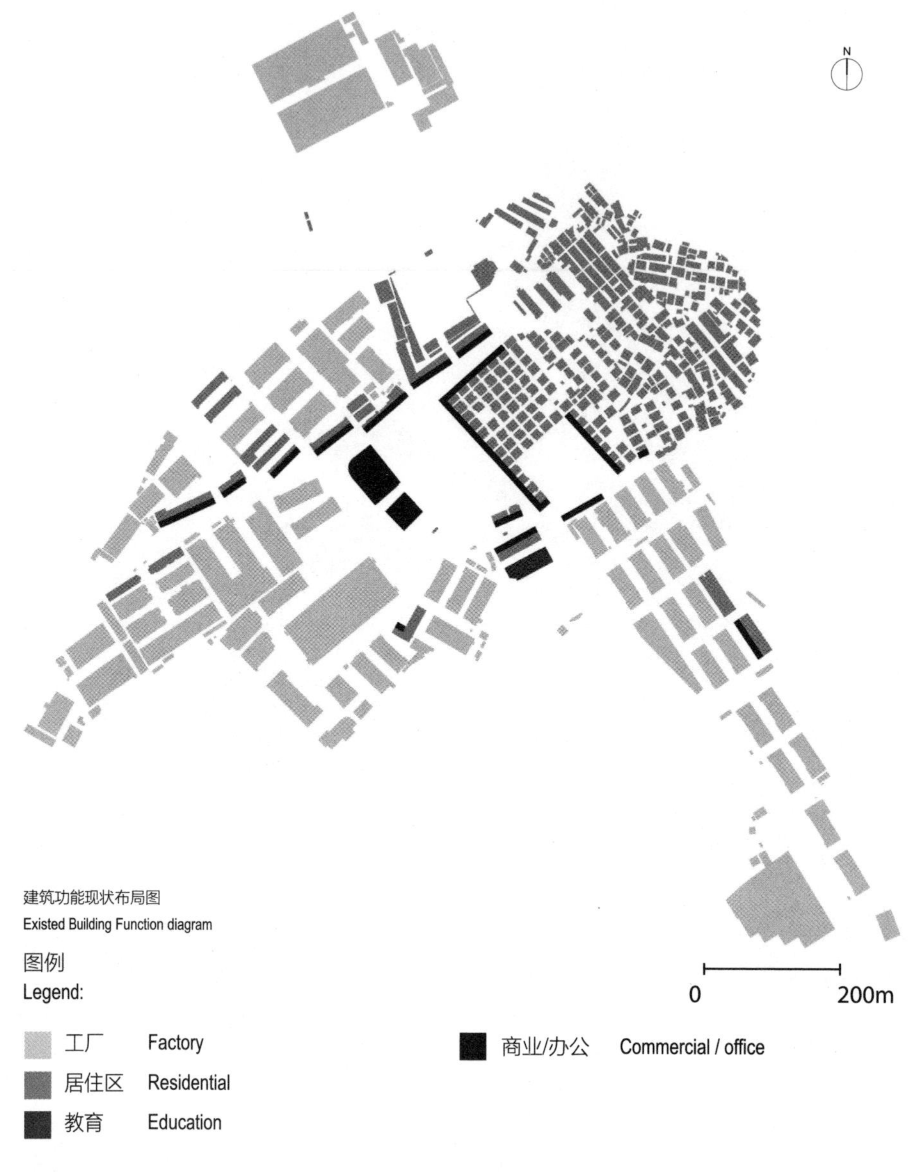

建筑功能现状布局图
Existed Building Function diagram

图例
Legend:

工厂 Factory
居住区 Residential
教育 Education
商业/办公 Commercial / office

建筑功能

村的西面和南面目前主要被工厂占据，其间零星分布着一些住宅楼，但大部分住宅功能集中在东北部。沿麻磡路、南麻磡路和公园东面可以看到商业核心，最大的超市就在公园旁边。该村的教育设施为一所幼儿园。

Building Use

The west and the south of the village are currently occupied mainly by factories with a few residential buildings scattered in between, but most of the residential functions are concentrated in the north-east. A commercial core can be observed along Makan Road, South Makan Road and east of the park, with the biggest supermarket being next to the park. The village also has a kindergarten as an educational facility.

建筑类型现状图
Existed build typology diagram

建筑类型

该村建筑有三种主要类型：板式厂房，板式住宅楼和方形点式住宅楼，比较旧的部分还有各种其他类型的住宅建筑。当然，也有一些特殊类型建筑物，例如最大的超市和一些特殊的工厂及办公建筑。

Typology

The village is characterized by three major typologies: long factory buildings, long residential buildings and square-shaped residential buildings. The older part of the village also has various other types of residential buildings. Of course there are also some buildings that have a special typology, for example the biggest super market or some special factories and offices.

沿着主街的长条住宅楼
Long residential buildings along the main street

方形住宅楼
Square-shaped residential buildings

麻磡村最大的商场
The biggest shop of the village

麻磡村里的典型工厂
Typical factories of the village

方形住宅楼
Square-shaped residential buildings

方形住宅楼
Square-shaped residential buildings

麻磡村其他住宅
The residence of the village

麻磡村其他住宅
The residence of the village

4.4 规划愿景
PLANNING VISION

愿景：双核心

整体概念的实质是：以连贯统一的空间思想为指导形成两个核心区域，并以此带动城市转型。这两个核心分别是：现有的公园，作为整个村庄的中央开放绿地；传统定居点内新改造的广场，作为对直接邻近居民的开放空间、园艺空间。这两个核心的总体改造思路接近，将导致核心逐步整合。

Vision: Dual Core

The essence of the overall concept is the aimed urban transformation led by a coherent, uniting spatial idea in the spirit of a programmatic and spatial dual core. The term *dual core* refers herein to the future two centers of the settlement – the existing park, as the central open green space for the whole village, and the newly transformed square within the traditional settlement, which would first and foremost serve as the open space for the direct neighbouring residents, respectively as the gardening space for those who are in charge of it, and subsequently as a programmatic focal point of the settlement in a wider sense. The direct proximity of these two cores, as well as their overall idea, would lead to a step-by-step unification.

总平面图

Masterplan

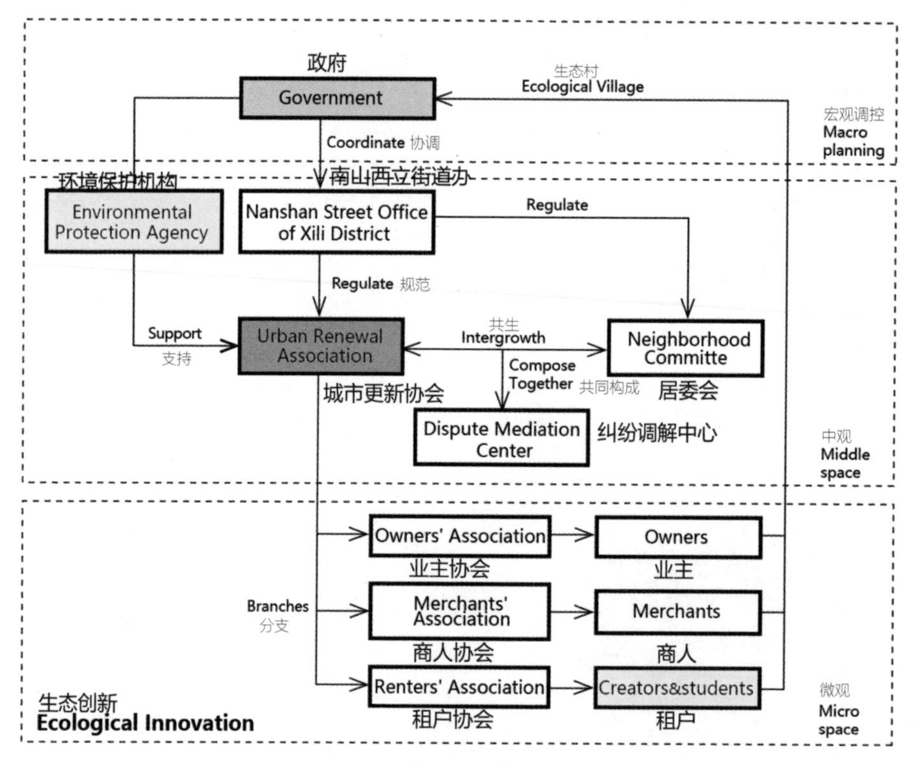

麻磡村运行机制图
Makan Village renewal Operation digram

运行机制

村庄应该变得高度整合化。每个部分都有自己的角色，可以在麻磡村相互依赖的网络中发挥作用。当然还有与深圳其他地区的联系，但在大多数情况下，它应该成为一个自给自足的生态系统。规划目标是创建一个创造性的可持续发展的示范村，是一个技术新颖的、能自给自足生活的游乐场。在这里，新想法可以在几天内快速实现，制作原型并进行部署/测试，只需在几步之遥的空间内便可完成。这是一个进步的企业家和技术工人聚集在一起，创造未来生活的地方。

Mechanism

The village should become a more integrated one. Each part has its own role to play in the interdependent network of Makan Village. Sure there is also the connection to the rest of the city of Shenzhen, but for the most part, it should become a self-reliant ecosystem. The goal is to create a model village of creative sustainability. A playground for novel technics and self-sufficient living. New Ideas can quickly developed, prototyped and deployed/tested within days, by just walking a view meters. It's a place where progressive entrepreneurs and skilled workers come together and create the living conditions of the future.

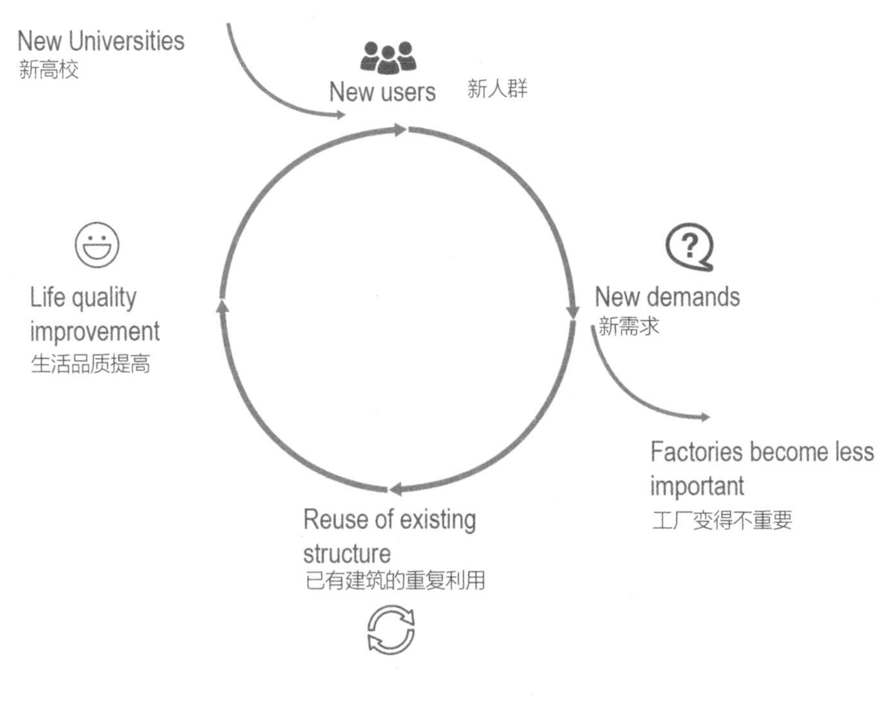

实施路径图
Implementation Approach

实施方案

分析表明，未来的麻磡村周围的社会和经济结构将在改造后发生根本性的变化。新建大学将为这个地区引入数千人，工厂将变得不那么重要。 这些变化肯定会影响这个区域。以下项目将展示麻磡如何适应未来发展，以及如何实现村落的多元化。为实现这一愿景，保留村落结构和建筑环境，通过新功能重新焕发区域活力。新旧结合将有助于应对未来的挑战，同时赋予村落全新的面貌。

Implementation

As the analysis shows, the future will bring radical changes to the social and economical structures surrounding Makan Village. The new universities will bring thousands of new people into the area and factories will become less important. These changes are sure to affect the valley.

The following projects will show how Makan will get fit for this future and how the vision of a village of mixed uses and a mixed society can become reality.

To achieve this vision, the structure of the physical and built environment will be preserved but revitalized with new functions. The combination of new and old will help to address the challenges of the future and help to generate an identity and a story for the village.

4.5 规划策略
PLANNING STRATEGIES

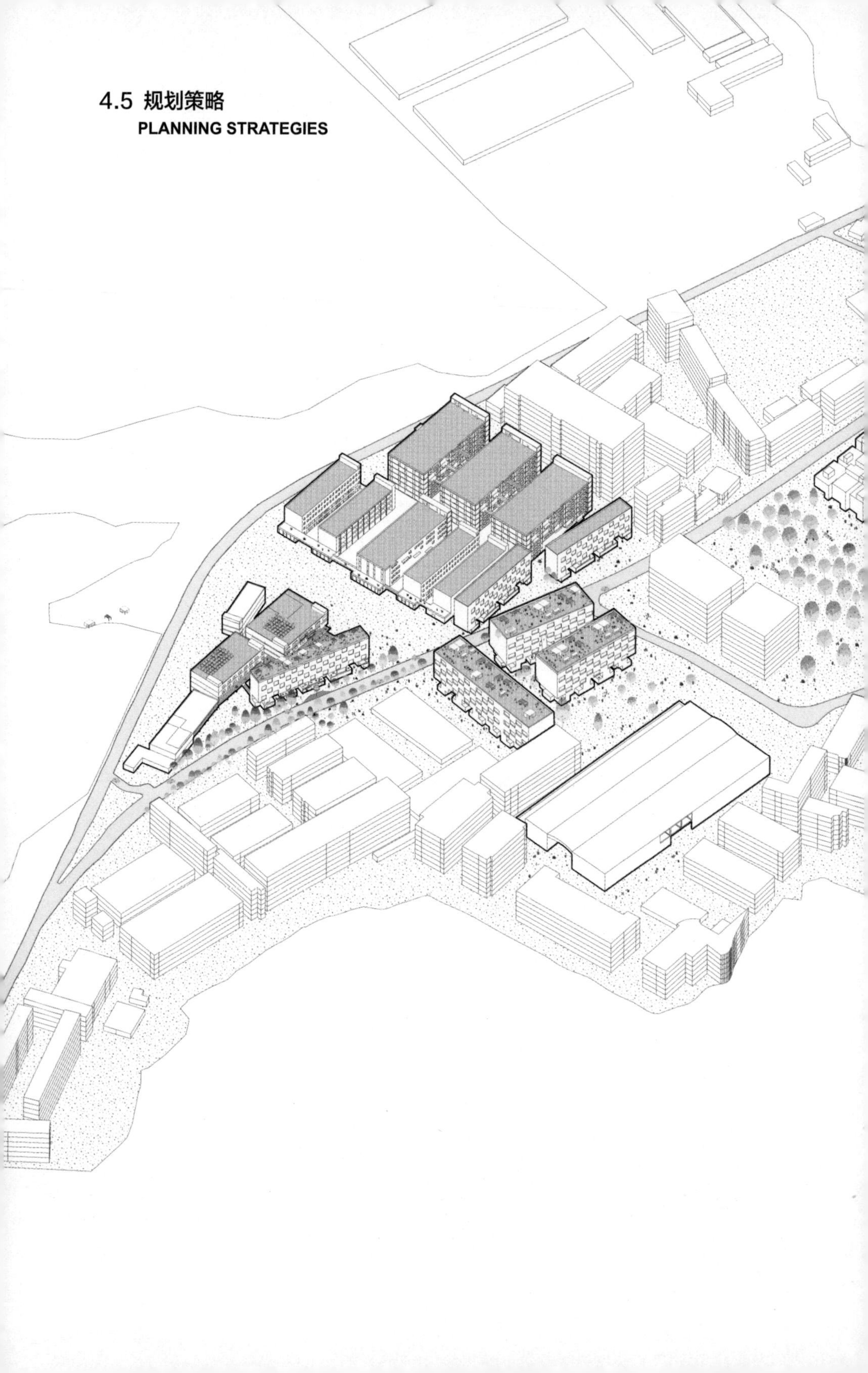

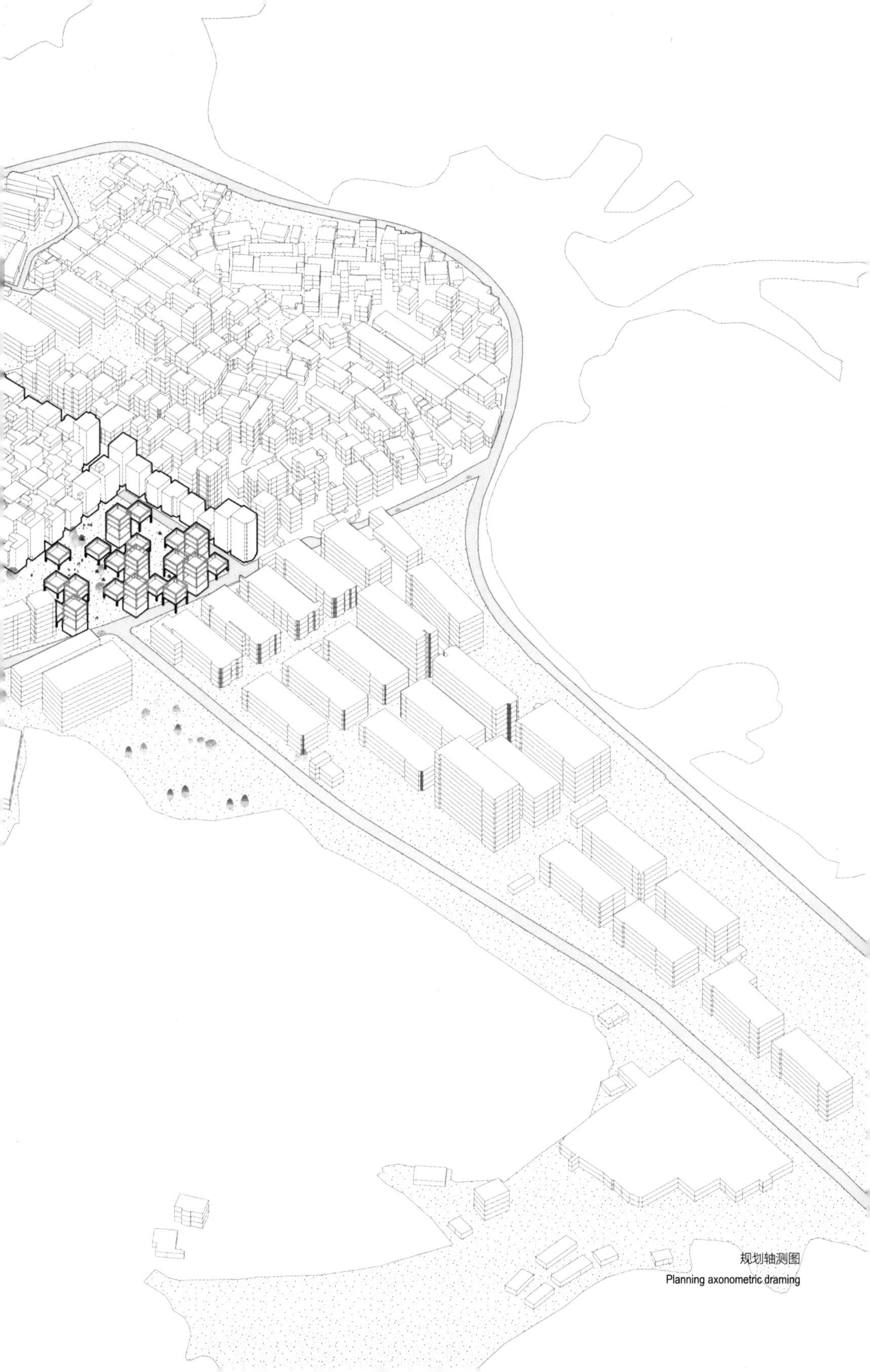

规划轴测图

Planning axonometric draming

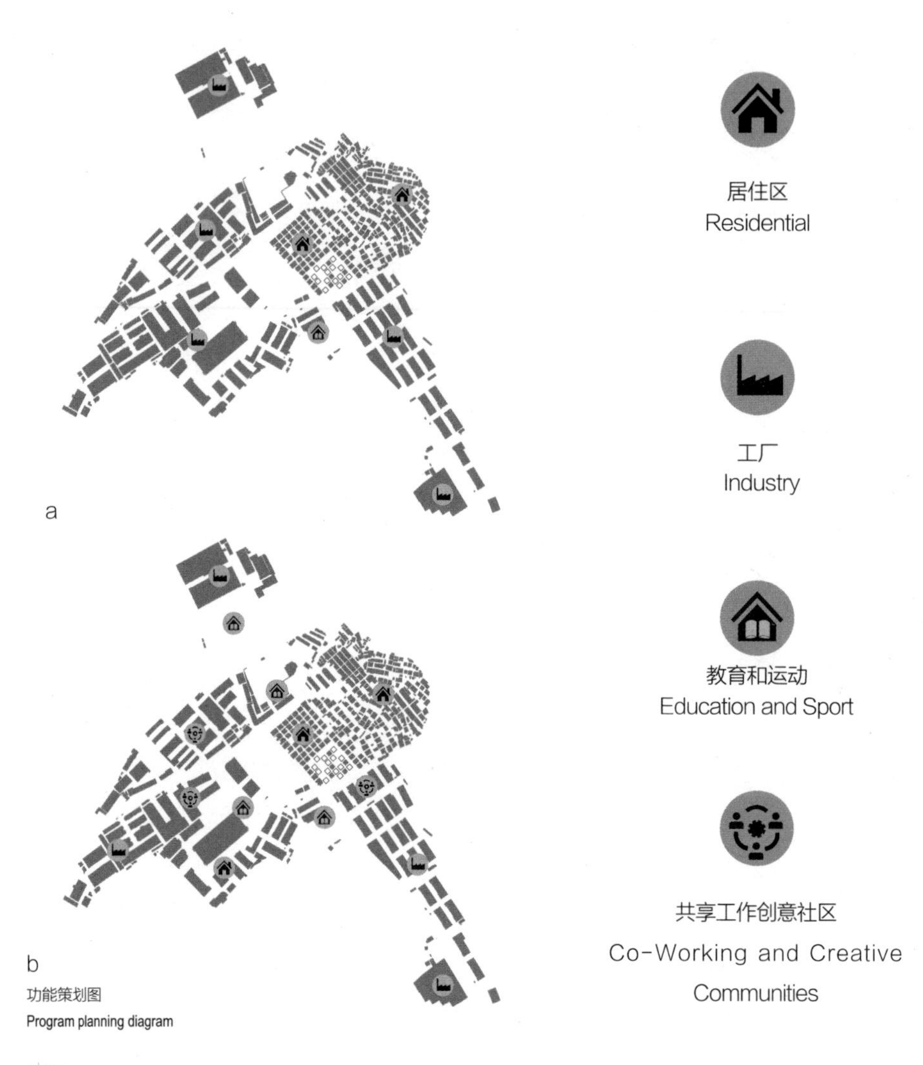

功能策划图
Program planning diagram

愿景

上图显示了地图的区域愿景。 地图 a 显示了目前村里的功能是如何扩散的。 地图 b 显示了未来的用途以及他们在麻磡村的位置。重要的是要看到现有的用途不仅要重新定位，而且要实现新的用途。 这些新用途将有助于解决村内及周围的未来发展问题。与此同时，重要的是麻磡村的未来不能完全消除旧用途。 这应该在保护了历史的同时，满足了未来村庄的发展需求。

Vision

The illustrations above are the vision for the area shown as a map. The upper map shows how uses are spread in the village at the moment. The map below shows the future uses and how they are located in Makan Village.It is important to see how not only existing uses are relocated but that new uses are implemented. These new uses will help to address the future development in and around the village.At the same time it is important for the future of Makan Village not to eradicate old uses in total. This should guarantee a development for the village where the (built) history meets the requirements of the future.

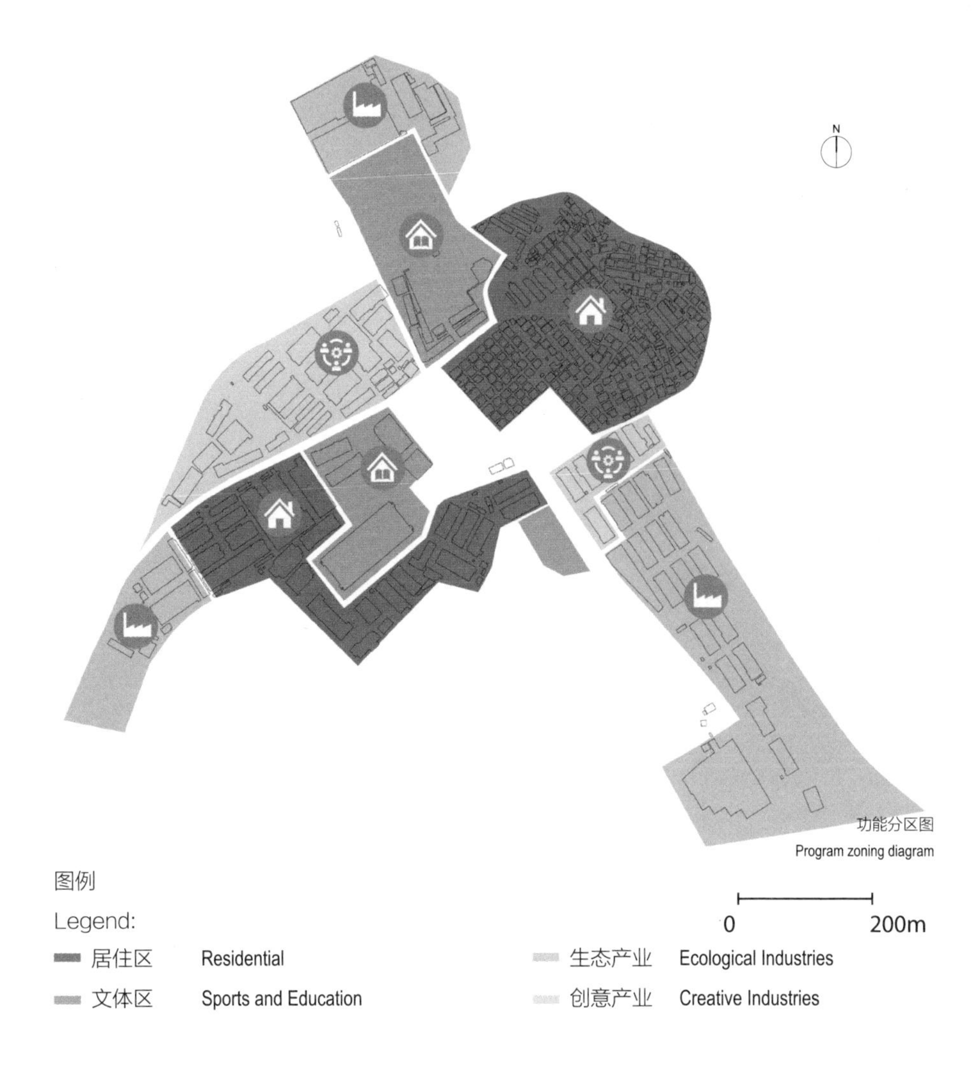

功能分区图
Program zoning diagram

功能分区

麻磡村目前建设用地以工业用地和住宅用地为主。村庄进行改造更新，应保持该地区的历史特征，同时推动其转型。北部的历史住宅区和南部混合功能的建筑作为住宅区，因此我们可以预测到麻磡村未来有更多居民，带来更多的需求及发展潜力。随着居民人数的增加，对体育和教育设施的需求也将会增加。规划新增的体育及教育设施会位于村庄的中心以及旧住宅区的西侧。西北部的旧工业区将转变为创意产业区，东南部地区的第一排厂房也将如此。

Zoning

Makan Village is currently dominated by industry and residence. Villages are soft-updated to maintain the historical character of the area while promoting its transformation. The historical residential area in the north and the mixed-use buildings in the south are used as residence, so we can predict that there will be more residents in the future, which will bring more demand and development potential. As the number of residents increases, so will the demand for sports and educational facilities. The planned new sports and educational facilities will be located in the center of the village and on the west side of the old residential area. The old industrial area in the northwest will be transformed into a creative industrial area, as will the first row of plants in the southeast.

发展轴线图
Development axis diagram

图例
Legend:
发展驱动 Development Drive

发展契机

作为发展的驱动力量，商业化的村庄道路也是游客最先接触的地方，因此应被视为迈向重塑麻磡村的第一步。麻磡路和麻磡南路的更新将带动村庄周围的发展，并以此为核心逐步发展至整个规划区域。

Trigger of Development

As the trigger of development we see the already very commercial and actively used main road of the village. This is the first place that visitors see and therefore should be seen as a first step towards a renewed Makan Village. The renewal along Makan Road and South Makan Road will trigger developments all around the village and this will gradually spread and conclude into a soft renewal of the whole planning area.

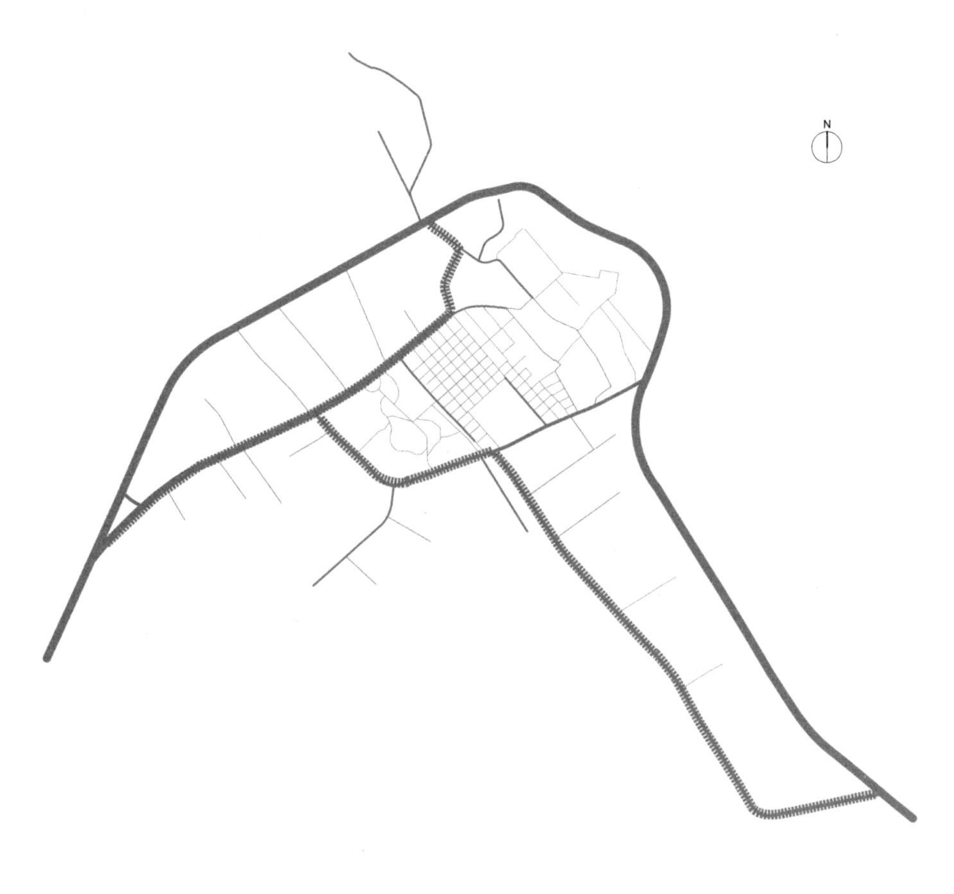

道路交通规划图
Road system plan

图例
Legend:
主要道路 Main Road

0 200m

道路系统

机动车交通可以通过村落的两条主街到达任意区域。两条主街不仅连接村落内部单一区域，同时也是进出的门户。其他街道则仅用于公共交通、后勤运输、自行车和人行步道。

Road System Concept

These two main streets through the village make all areas accessible for private motorised transportation. These two roads do not only connect the single areas within the village, they are also the gateway in and out. Other streets will be used for public transportation, deliveries, bicycles and pedestrians only.

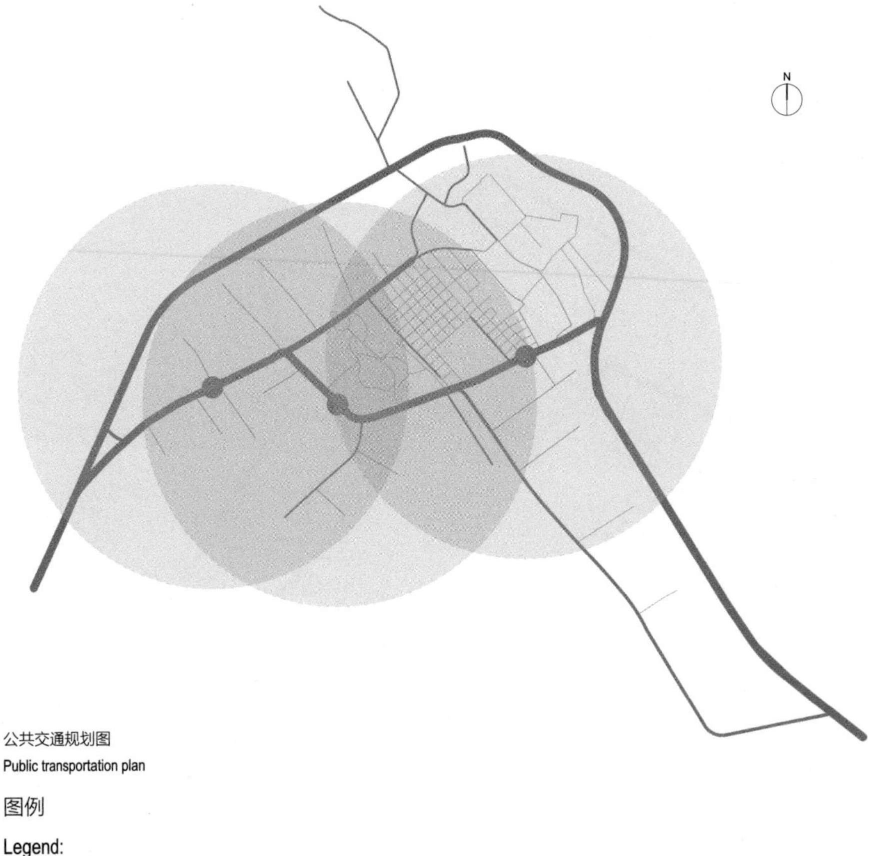

公共交通规划图
Public transportation plan

图例

Legend:

公交路线	Bus Line
300m 辐射圈	300m Radius

0 200m

公共交通系统

公交系统的微小调整就可以将新的开发站点更好地连接到公共交通网络，并有助于为村落提供更多的服务。 公共汽车的路线保持不变，因为它已经通过了村庄最重要的部分。 公交站稍稍移动使得它能更好地覆盖 300m 半径范围内的村庄， 这是步行时间不到 5 分钟的距离。 公交系统将来会连接到深圳的地铁网络。

Public Transportation Concept

Small changes in the bus system connect the new development sites better to the public transportation network and help serving a bigger share of the village. The route of the bus is left as it is, because it already passes the most important parts of the village. The bus stops are moved a little bit, so they cover more of the village within a 300m radius. This equals a walking time of less than 5 minutes. The bus will be the connection to the subway network of Shenzhen in the future.

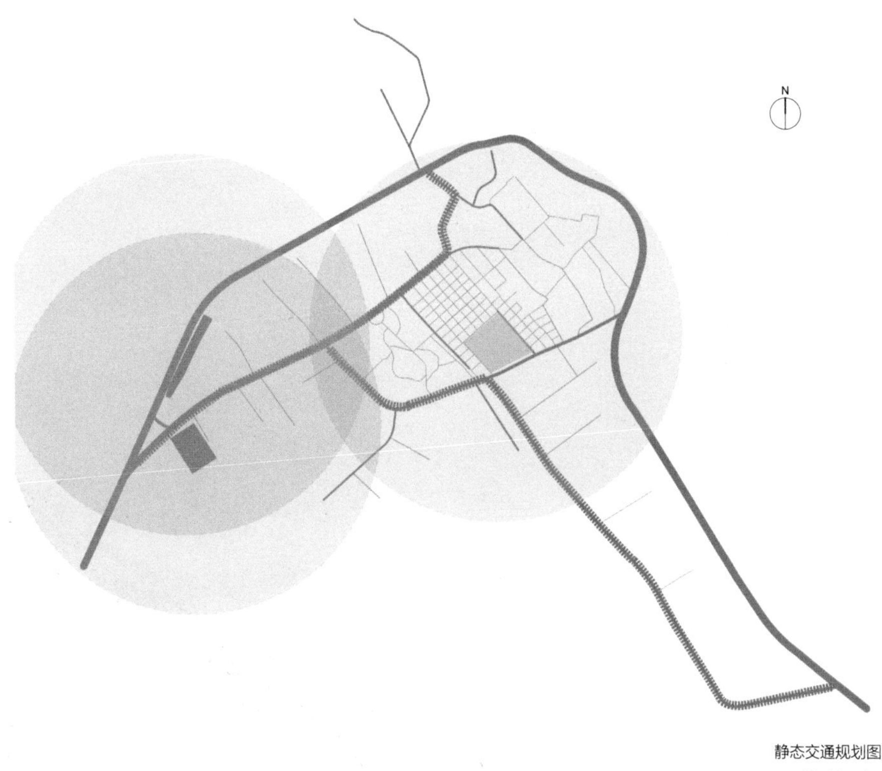

静态交通规划图
Panleing plan

图例

Legend:

主要道路 Main Road

300m 辐射圈 300 Meter Radius

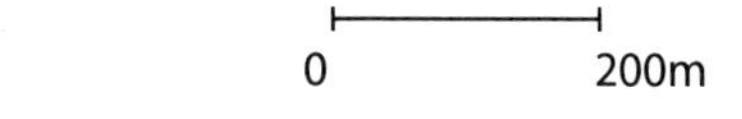

停车系统

规划的停车位主要集中在主要道路两边和集中停车场。能为其他的活动提供更多的公共空间。停车概念的关键点是在堆叠园艺场设置地下停车场。 其他停车位主要集中在村庄的入口处，从而减少了村内的车流。

Car Parking Concept

The new parking system concentrates on the main roads and tries to centralise the parking: this way more of the public space can be used for other activities.

One key point of the parking concept is the underground parking lot under the gardening site. The other parking spaces are located at the entrances to the village, which results in reduced car traffic inside the village.

图例

Legend:

- 公共活动空间
- 主要的步行系统
- 重要的步行系统
- 居民的步行系统
- 活力广场

慢行交通规划图
Pedestrain system plan

步行与活动空间

规划将村庄的核心区域扩展到公园之外，因此许多新的区域为行人开放，并且设计有品质的城市空间。品质核心连接了村庄的不同部分，使得核心更容易进入，并改善了村庄的整体步行性。新的活跃地面层模式反映了该村的新发展和扩展核心。

Pedestrians & Active Groundfloor Concept

The planning expands the core of the village beyond the park, as many new areas are opened up for pedestrians and are designed with qualitative urban space. This qualitative core connects the different parts of the village that are opened up, making the core more accessible and improving the overall walkability of the village. The new active groundfloor pattern reflects the new developments and the new expanded core of the village.

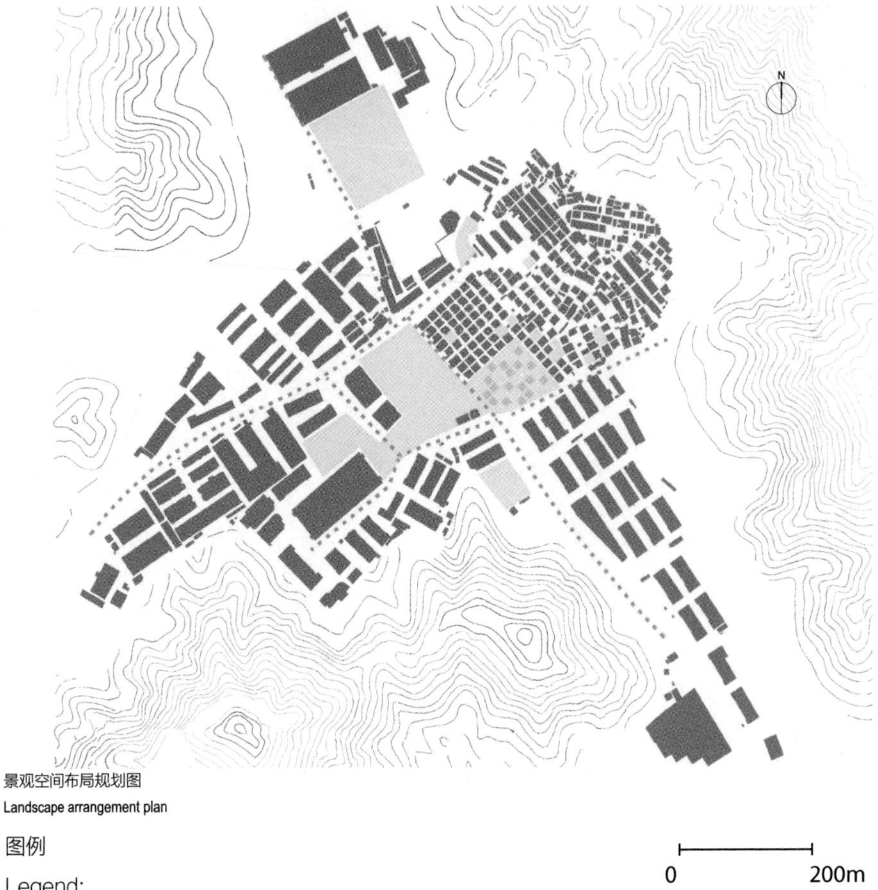

景观空间布局规划图

Landscape arrangement plan

图例

Legend:

(浅灰色块)	绿地空间	Green Space
(虚线)	绿轴	Green Axis
(等高线)	高度线	Height Lines

绿地及绿轴

麻磡村是自然保护区的一部分，被大自然和丘陵所环绕，绿色空间是城市更新规划概念的主题之一。拥有着双核心，麻磡的中心是一个大型的开放式绿色空间，每个人都可以使用。小块的绿地分散在村庄周围，是更多的半开放空间，居民可以在较私密的环境中享受绿色空间。这些空间都与绿色轴的双核相连，为村庄提供了清晰的绿色概念。

Green Spaces and Axes Concept

As Makan Village is part of the nature protection area and therefore surrounded by nature and hills, the green space is one of the main topics of the concept. With the dual core, Makan Village's center is a big open green space, which is accessible to everyone. The smaller green spaces, which are scattered all around the village, are more semi-public spaces, where the residents of the neighbourhood can enjoy green space in a more private setting. These spaces are all connected to the dual core with the green axes, giving the village a clear green concept.

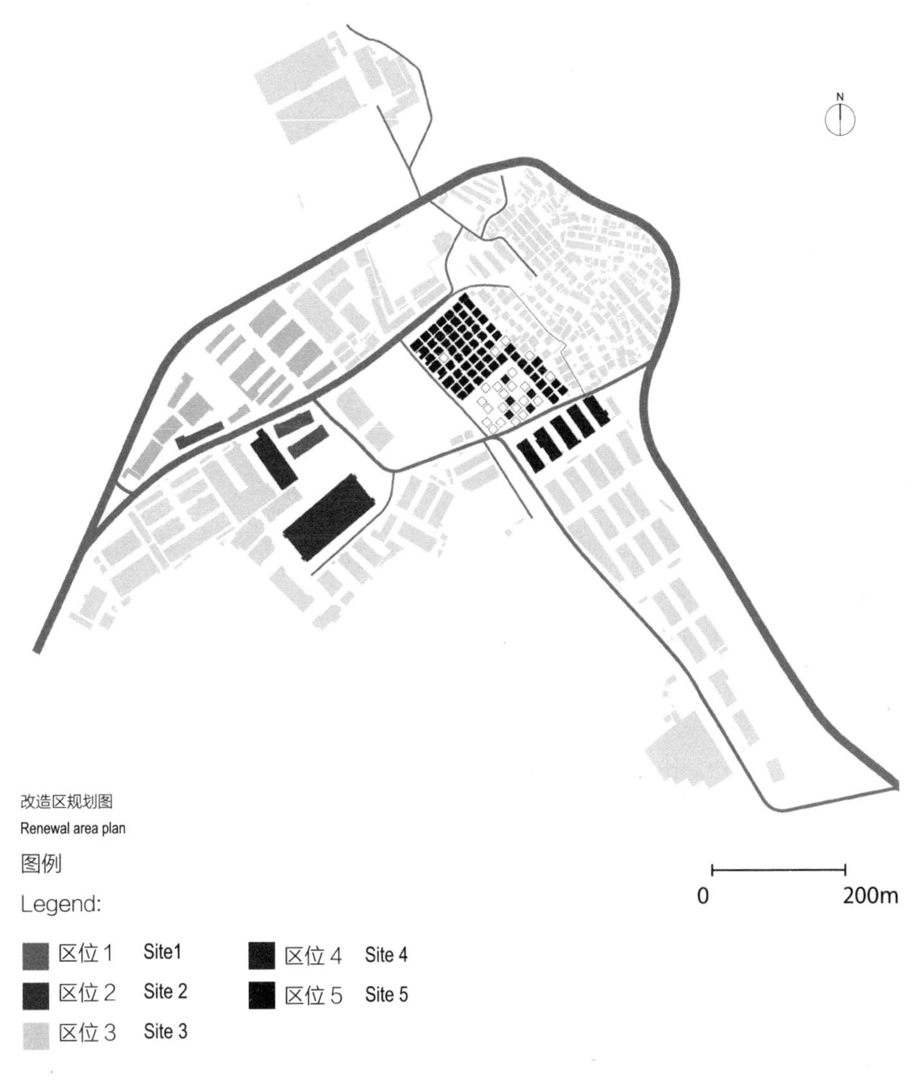

改造区规划图
Renewal area plan

开发区域

如图所示，是对麻磡村的空间结构进行干预的整体尝试。所有单一概念都表现为一个相关的相互影响的综合体。综合体会导致解决方案的转变——无论是逐步发展，还是作为巨大的、单一的影响。

Development Sites

The graphic depicts the overall attempt to intervene within the urban structure of Makan Village. All of the singular concepts are shown as one coherent complex of linked intervention, which should lead to a transformation of the settlement – either as step-by-step-development, or as huge singular intervention.

常住在深圳城中村的外来人口占总人口90%以上，而深圳的平均年龄仅为32.5岁，深圳的城中村社区是一个属于青年人的社区。如何打造一个年轻、具有活力和创意的青年社区自然成为研究团队的主要设计任务。

Migrants take up 90%of the population in urban villages. Additionally, the average age of population in Shenzhen is around 32.5. Thus, urban villages are communities for the youth. How to construct a young, active and creative community for young people becomes the major task of the design.

5 麻磡村的改造实验——城中村更新
青年活力社区营造
MAKAN RENOVATION STUDY_URBAN VILLAGES
CREATING OF YOUTH COMMUNITY

5.1 麻磡村改造
MAKAN VILLAGE RECONSTRUCTION

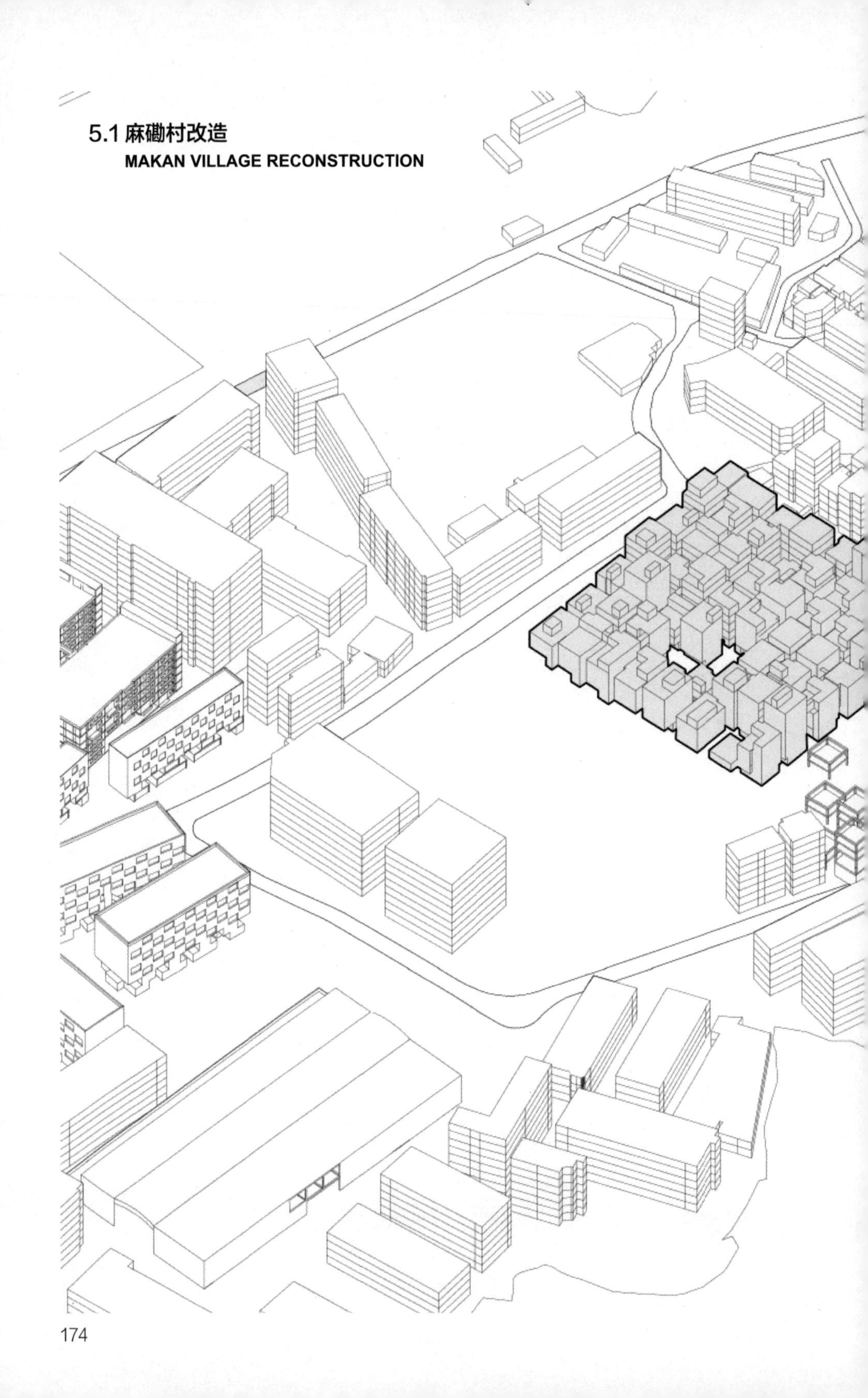

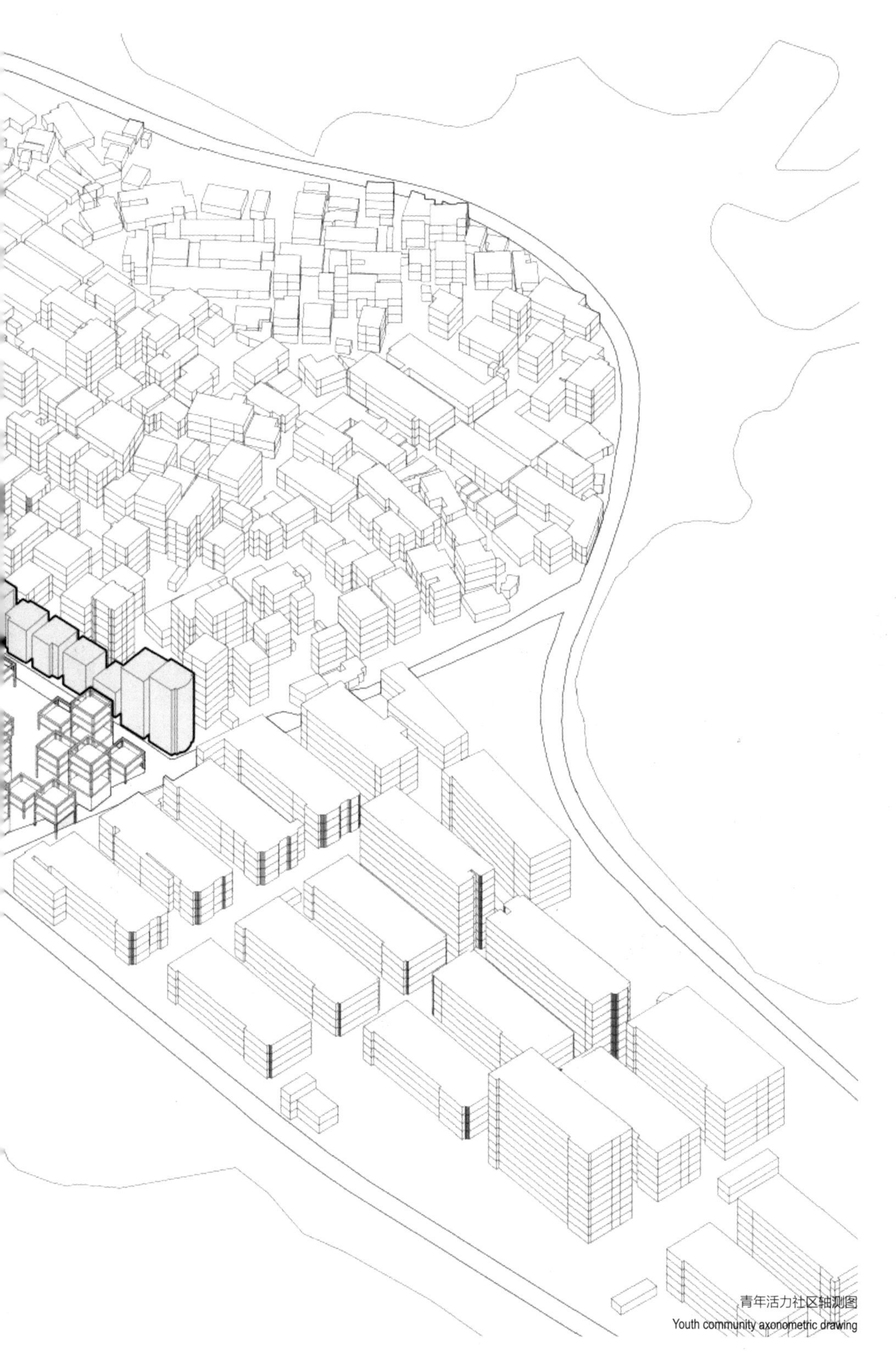

青年活力社区轴测图
Youth community axonometric drawing

Calumet Farm
Lexington, Ky.

青年社区透视图

Youth community perspective

功能策划图

Program planning diagram

新影响

由于麻磡村的未来发展需求，现有的住宅结构将被不同的新旧功能和用途所包围。这种发展模式将对场地本身产生影响。由于麻磡村的改造总体方案涉及现有功能和居民，因此这个中心区域必须符合现有环境和未来规划两个方面的要求。为了尽可能保留片区的城市特征，而该区域又必须有对外连接城市的重要功能，开发对公众开放的区域以及满足居民近距离需求的区域至关重要。

New Influences

Due to the future development in Makan Village, the older residential structures will be surrounded by different old and new functions and uses. This development will have an effect on the site itself. Since the overall draft for Makan Village includes old functions and residents into the new development, this central area has to fit the old as well as new requirements. The aim is to preserve as much of its character as possible.

At the same time the area will connect the important functions outside the site. For this reason it is critical to develop areas open to the public as well as areas that serve the needs of close by residents.

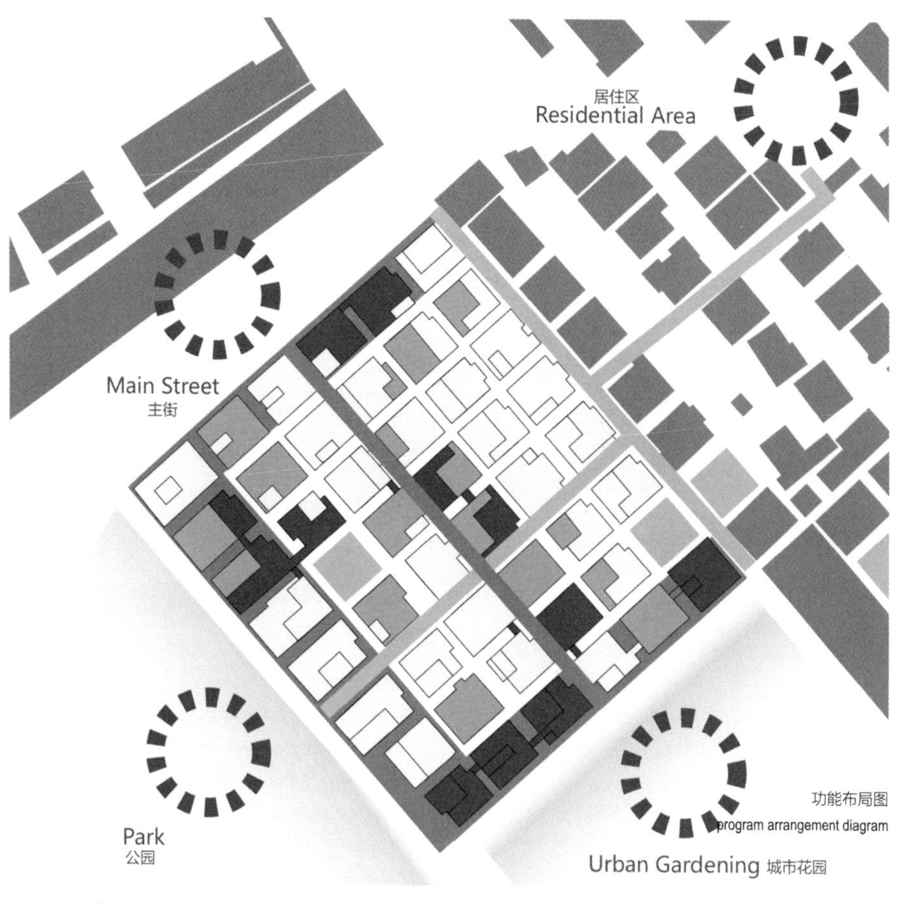

功能布局图
program arrangement diagram

居住道路 Residential Path
商业道路 Commercial Path
商业片区 Commercial Area
口袋公园 Pocket Park
屋顶绿地 Green Area (Rooftops)

总体规划

为了解决前面提到的挑战，基地的总体规划侧重于公共空间和没有围墙围合的私人空间之间的差异。这样，私人空间在起连接功能的同时仍为居民提供私密性。商业区将位于一楼和其他对公众开放的区域，包括临街面以及连接园艺区和村中心的一条街道。基地其余区域旨在满足居民的需求，并确保他们的隐私，包括小型口袋公园以及住宅区和大型绿地之间的联系空间。

Masterplanning

To address the former mentioned challanges the master plan for the site focuses on a difference between public and privat spaces without walls and fences. This way the area has a connecting function while still providing privacy for residents. Commercial areas will be located in the ground floor and in those areas that are opened to the public. This includes the borders as well as the street going through the center connecting the urban gardening area with the main street.The rest of the site is designed to fit the needs of residents and to ensure their privacy. This includes small pocket parks as well as connections between other residential areas and the big green spaces.

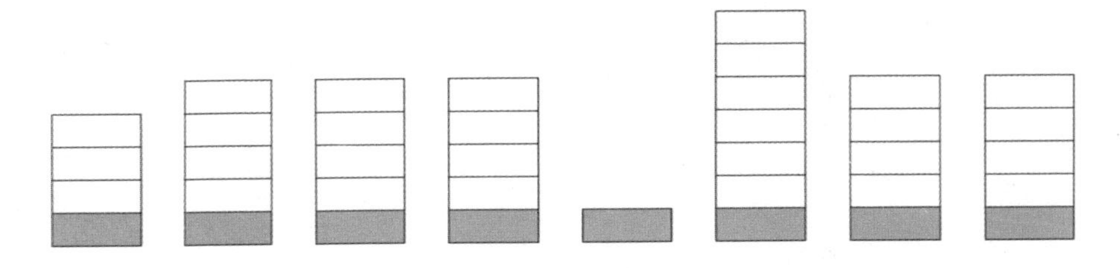

发展：第一步

先开发建筑物底层用于新型的商业。

Development: Step 1

At first only the ground floors of the single buildings will be developed with new commercial uses.

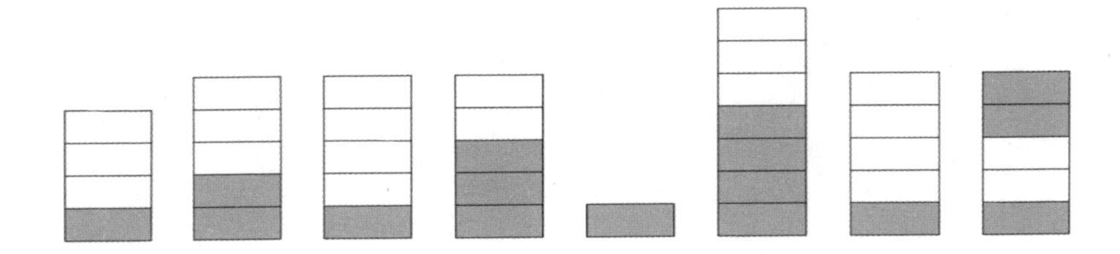

发展：第二步

房东可以决定房间的功能，例如民宿或餐厅。为了确保该区域不会失去所有居民，每栋建筑建议至少有 1/3 留作住宅用途。

Development: Step 2

Later on home owners can decide to implement new functions that require more space like hotels or restaurants on higher floors. To ensure that the area does not lose all its residents, at least one third of each building has to be reserved for residential use.

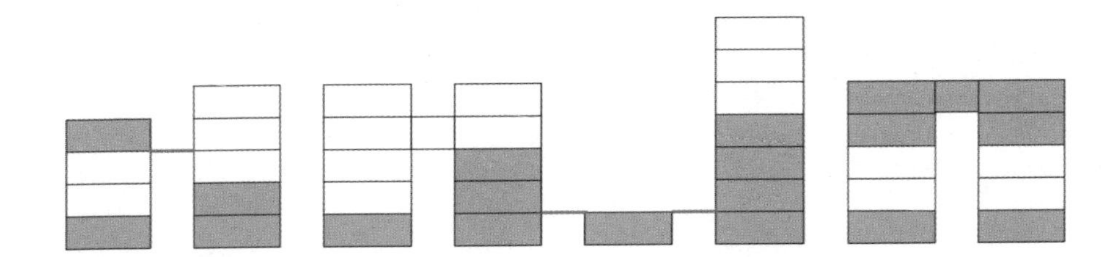

发展：第三步

单栋建筑可能会因为新的用途而被连接从而发展出新的功能。但建筑物不允许在底层连接，保证道路不会减少。预留至少 1/3 的住宅是十分必要的。

Development: Step 3

New functions spread and single buildings may be connected to compliment new uses. The buildings are not allowed to be connected on the ground floor, so paths are not lost. Still it is important that at least one third of the area is used for residential purposes.

实施路径图

Implementation approach

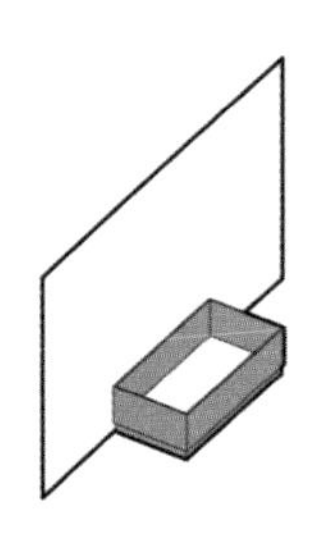

花园
Urban Gardening

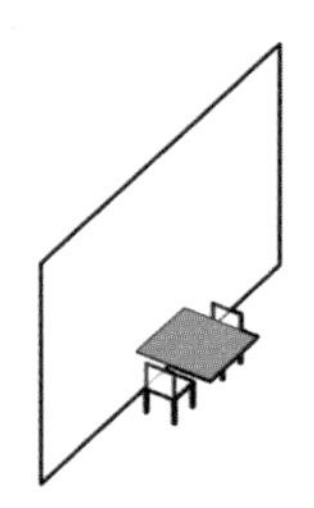

棋牌桌
Playing Cards

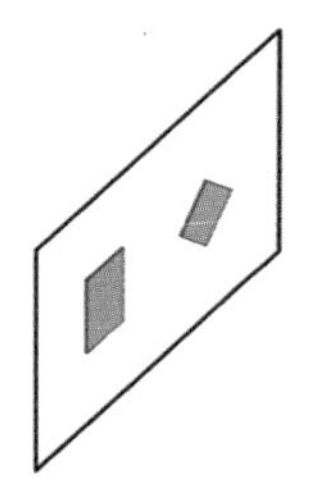

公共艺术
Public Art

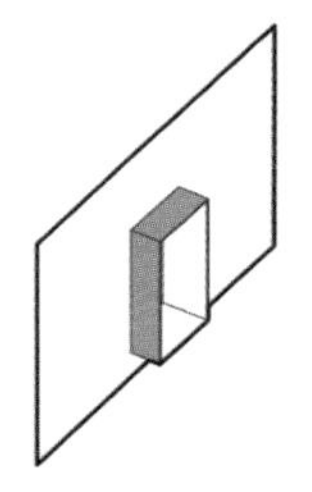

橱窗
Showcase

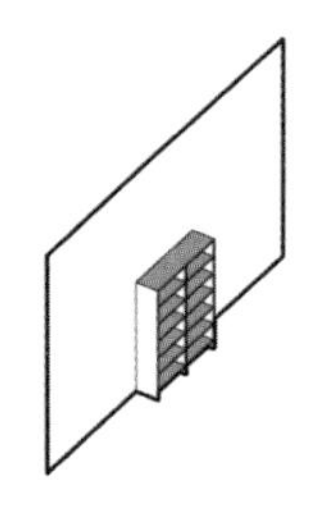

开放书架
Open Library

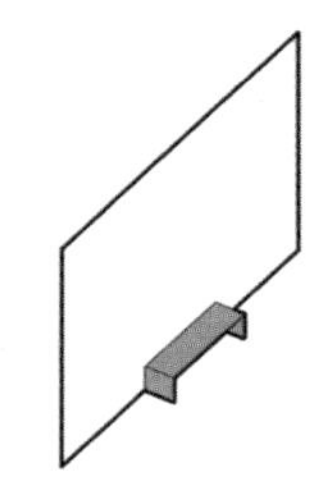

长凳
Bench

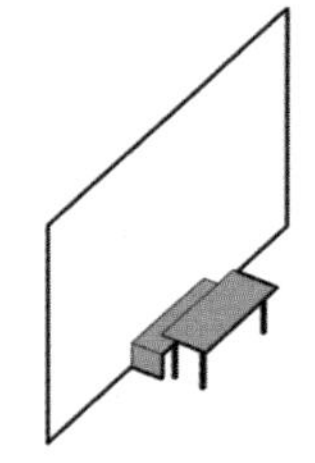

桌凳
Bench with Table

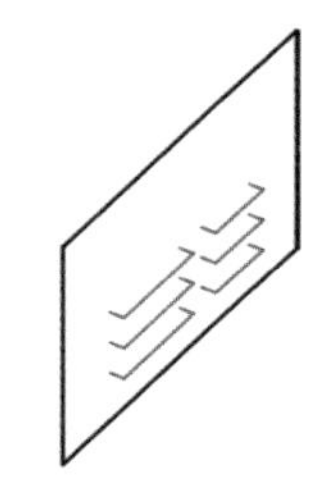

晾衣架
Clothes Drying

原则 1

住宅街道将会提供多样化的功能满足人们的需求，一些在建筑立面上实施，一些在街道空间上实施。 同时，建筑外墙的设计将确保居民有足够的隐私。

Principles 1

Residential streets will provide a series of functions suiting the needs of the inhabitants in the future, some implemented on the facade, some on the street. Facades will be designed in a way that ensures sufficient privacy for the residents.

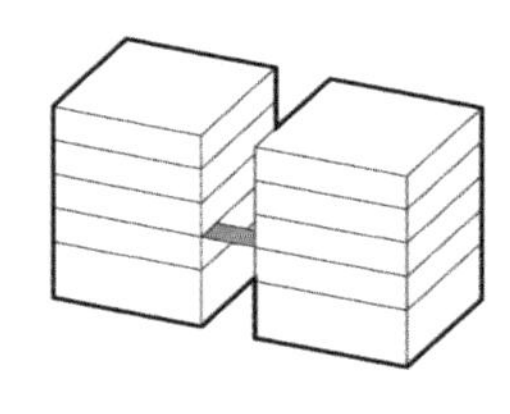

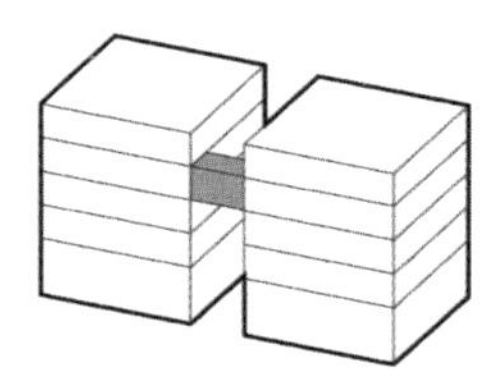

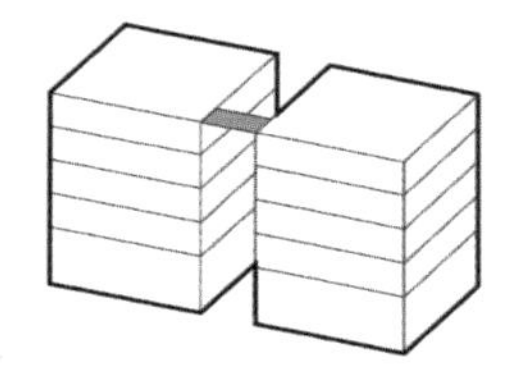

原则 2

建筑物可以以不同方式连接，从而使它的潜在用途种类最大化。 建筑之间的连接可以在不同层高上，也可以在不同的功能之间。

Principles 2

Buildings can be connected in different ways to maximize the variety of potential uses. Connections can be implemented on different levels and can include different functions.

改造策略图
Renewal strategles

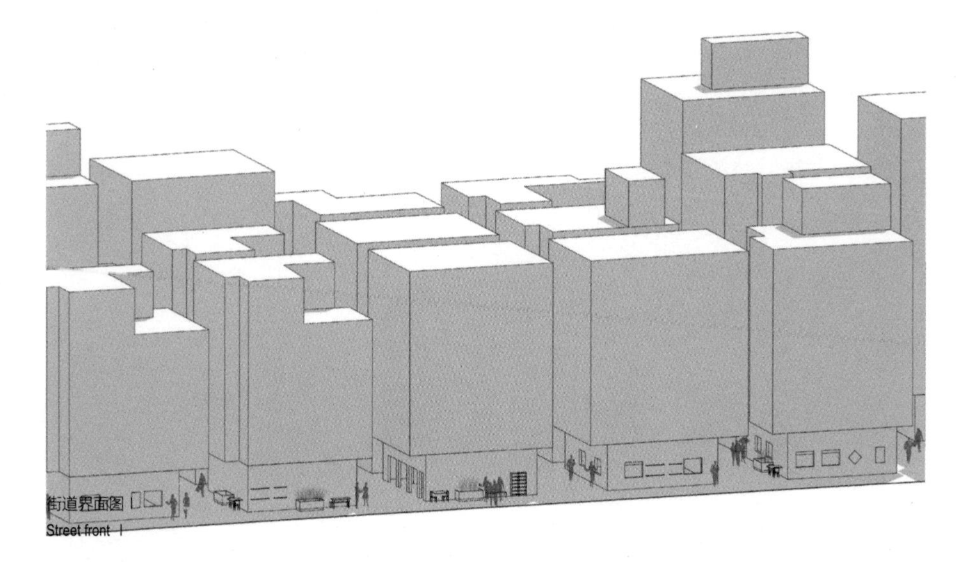
街道界面图 I
Street front I

居住街道

这展示了如何在街道空间中实施原则的范例。 主要目标是为居民和行人提供用途，并让他们聚集在这条街上，从而避免他们去更私密的区域。

Residential Street

This shows an example of how the principles can be implemented in the street space. The main goal is to provide uses for residents and people who walk through to concentrate them on this street and keep them out of more private areas.

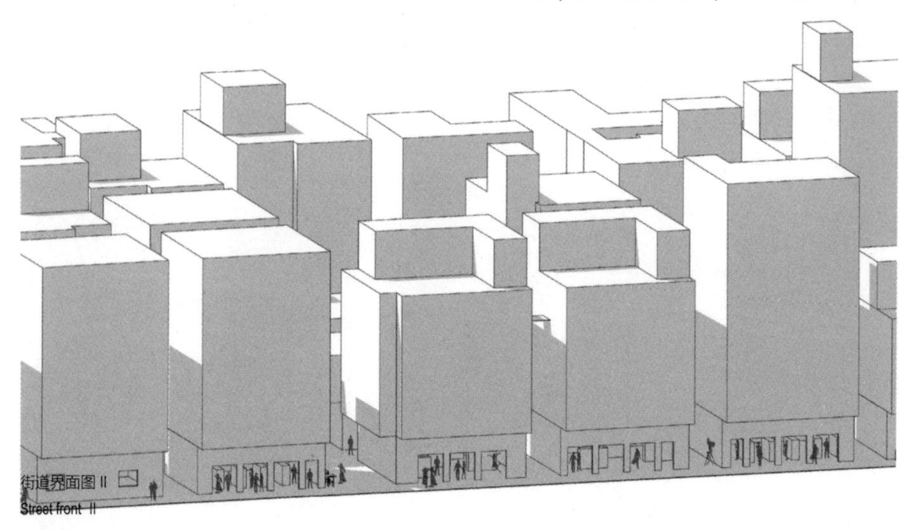
街道界面图 II
Street front II

商业街

商业街的外墙向公众打开，吸引人们进入。 商业街的重点不在于像居住街道一样，尽量满足当地居民的需求，商业街也意味着让人们远离更私密的区域。

Commercial Street

The facades of the commercial street open up to the public and invite people to pass through. The focus is not so much on the needs of local residents as in the residential streets. Commercial streets are also meant to keep people out of the more private areas.

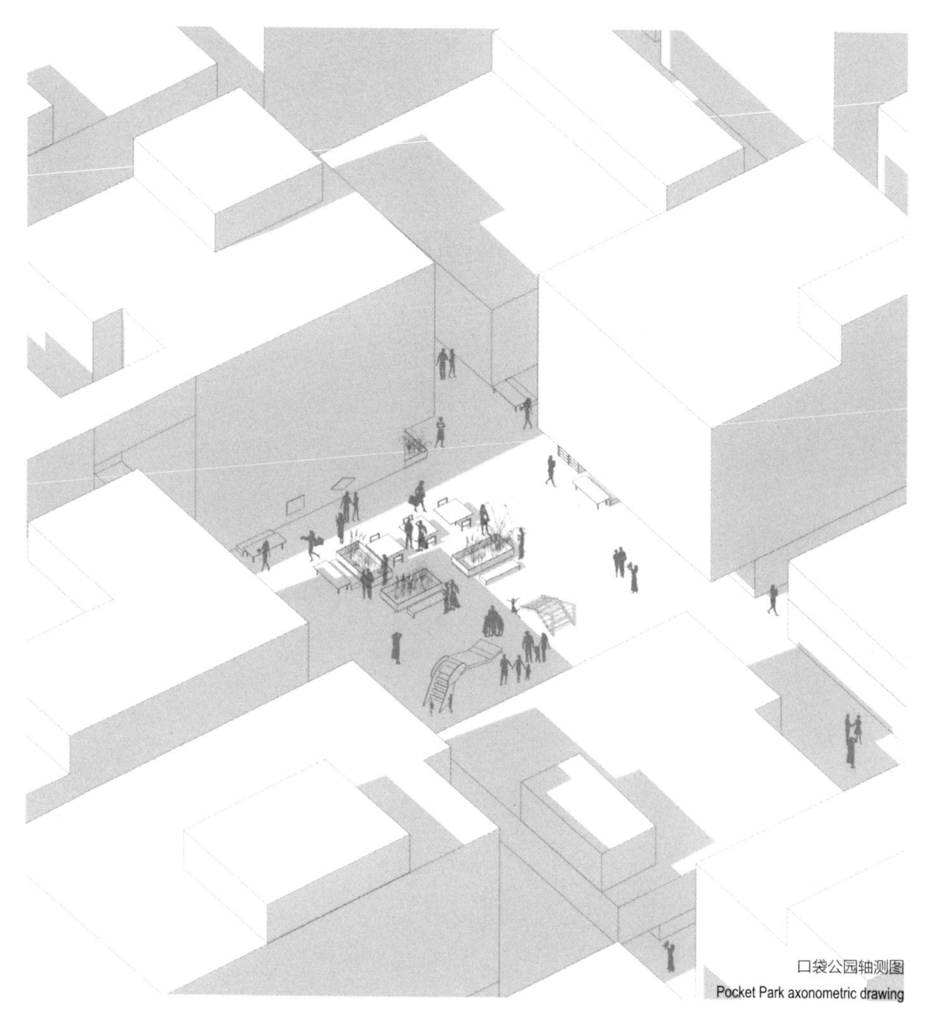

口袋公园轴测图
Pocket Park axonometric drawing

口袋公园

居住区域的口袋公园将满足居民的日常活动需求，居民的使用需求决定使用功能。公共客厅促进人们之间的交流。

Pocket Park

The small parks within the residential area will be used to meet the needs of the inhabitants. The functions will be decided together with the residents. This will create public living rooms suiting each community.

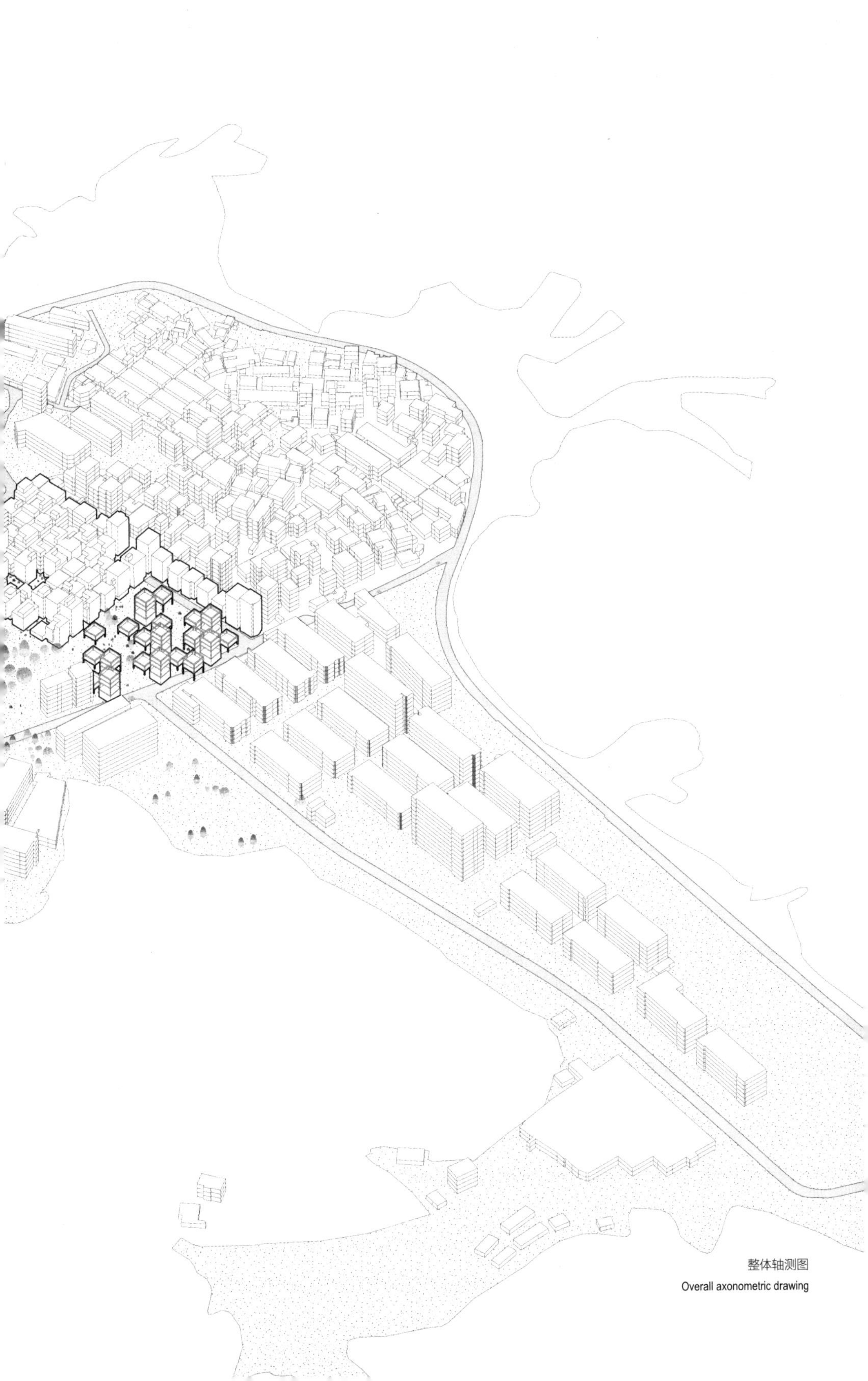

整体轴测图

Overall axonometric drawing

5.2 创意集市
CREATIVE MARKET

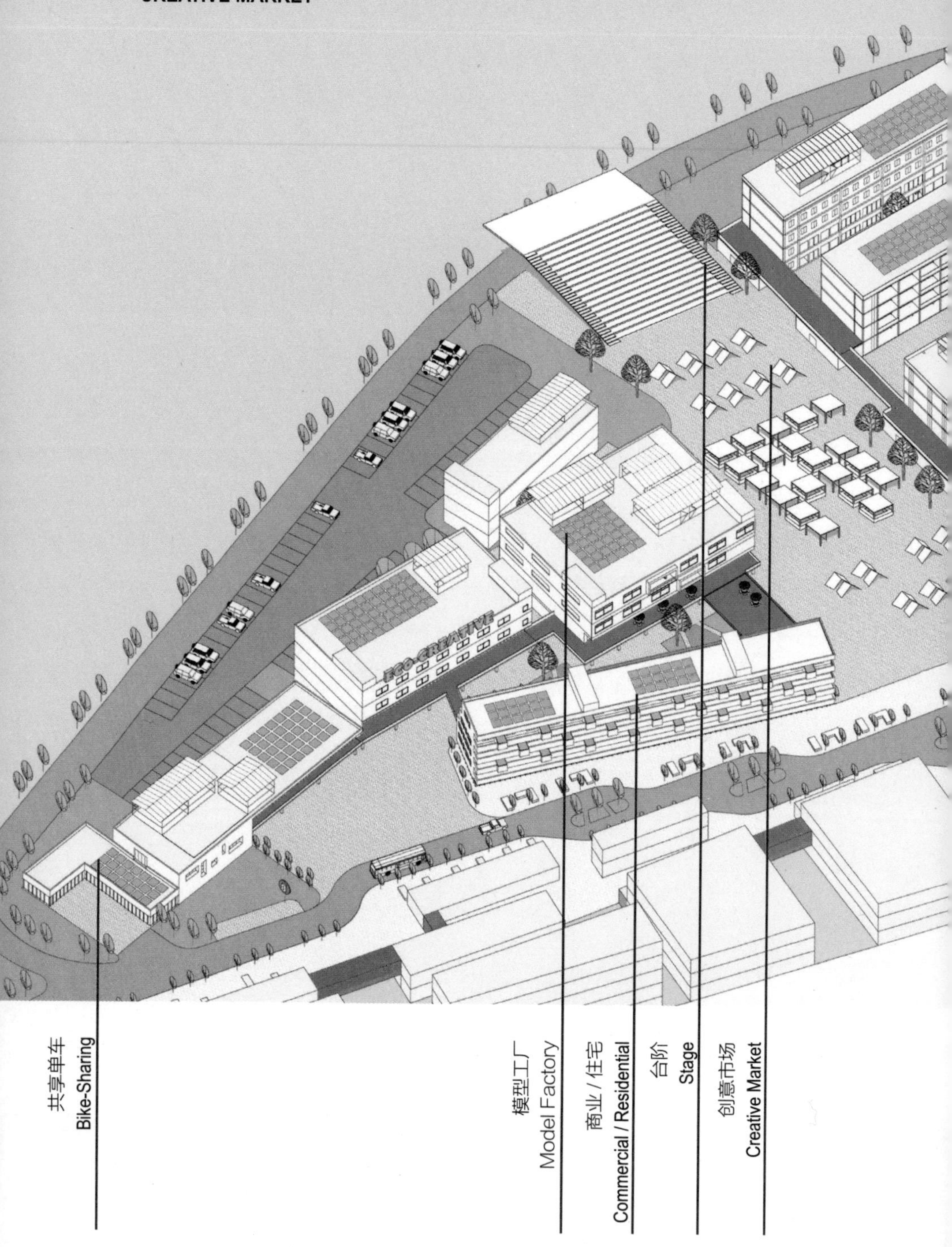

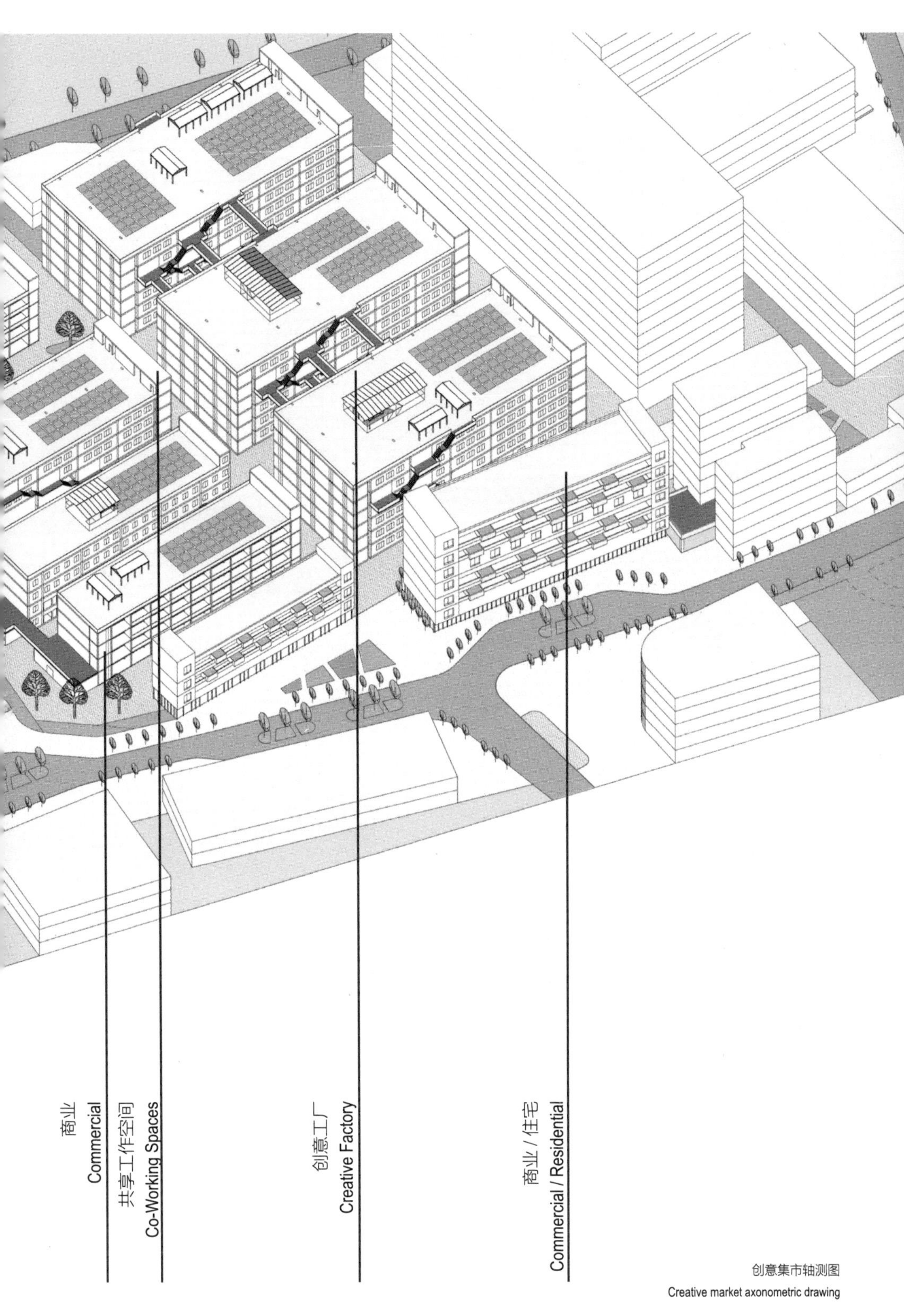

创意集市轴测图

Creative market axonometric drawing

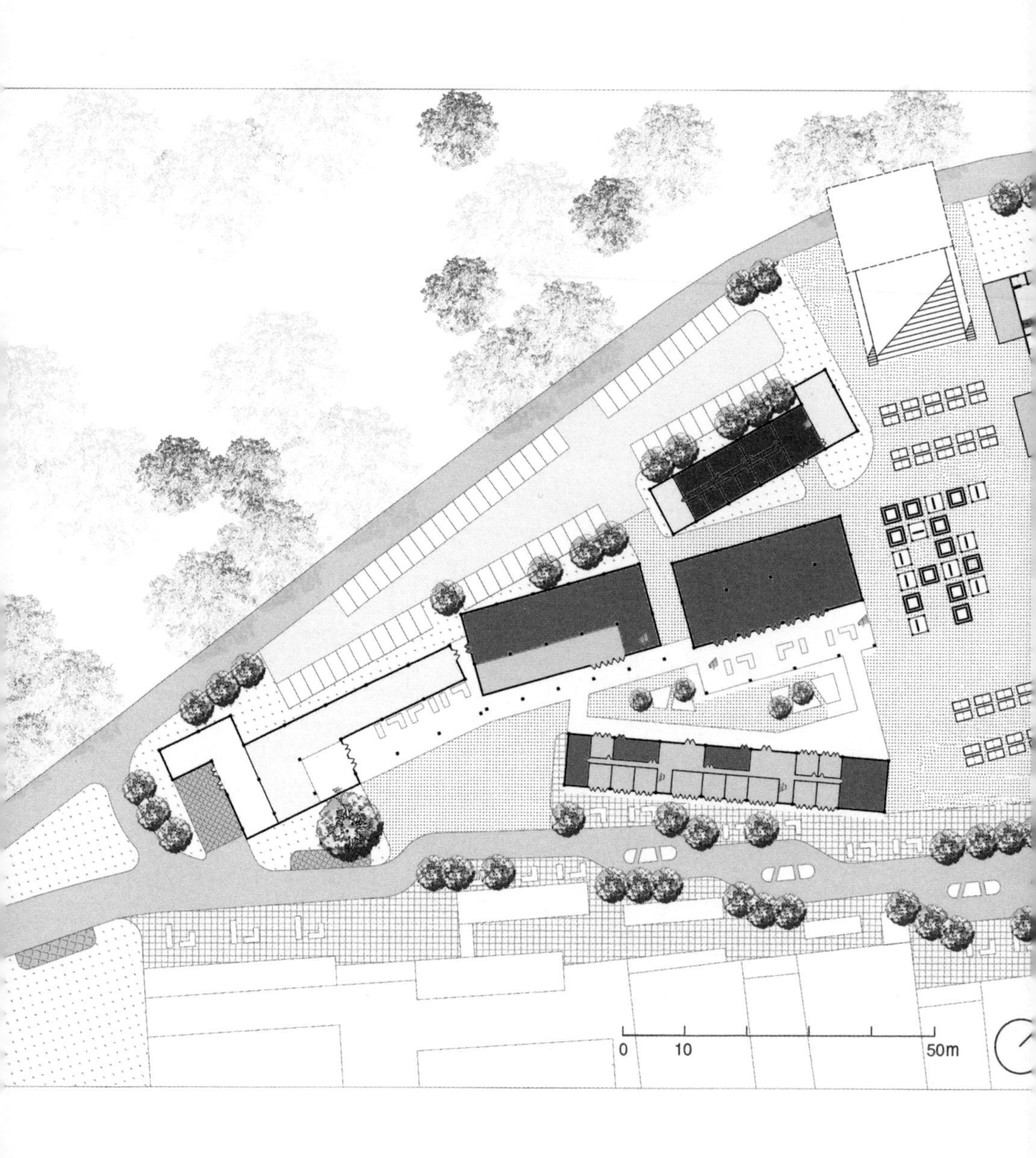

展览 Exhibition
生态毂 Eco-Hub
设备 Facilities

建筑功能布局图
Building function arrangement diagram

设计步骤

发展不是一蹴而就，而应随着时间的推移而发展。应确保有富余的空间，因为一个富有创造力的社区需要互动，相互了解并开始建立公司， 这需要时间与空间。

Step by Step

The development should not be transformed all at once but rather grow over time. This should ensure, that there is overcapacity in spaces. A community of creative minds need to interact, get to know each other and start building companies. This takes time and space.

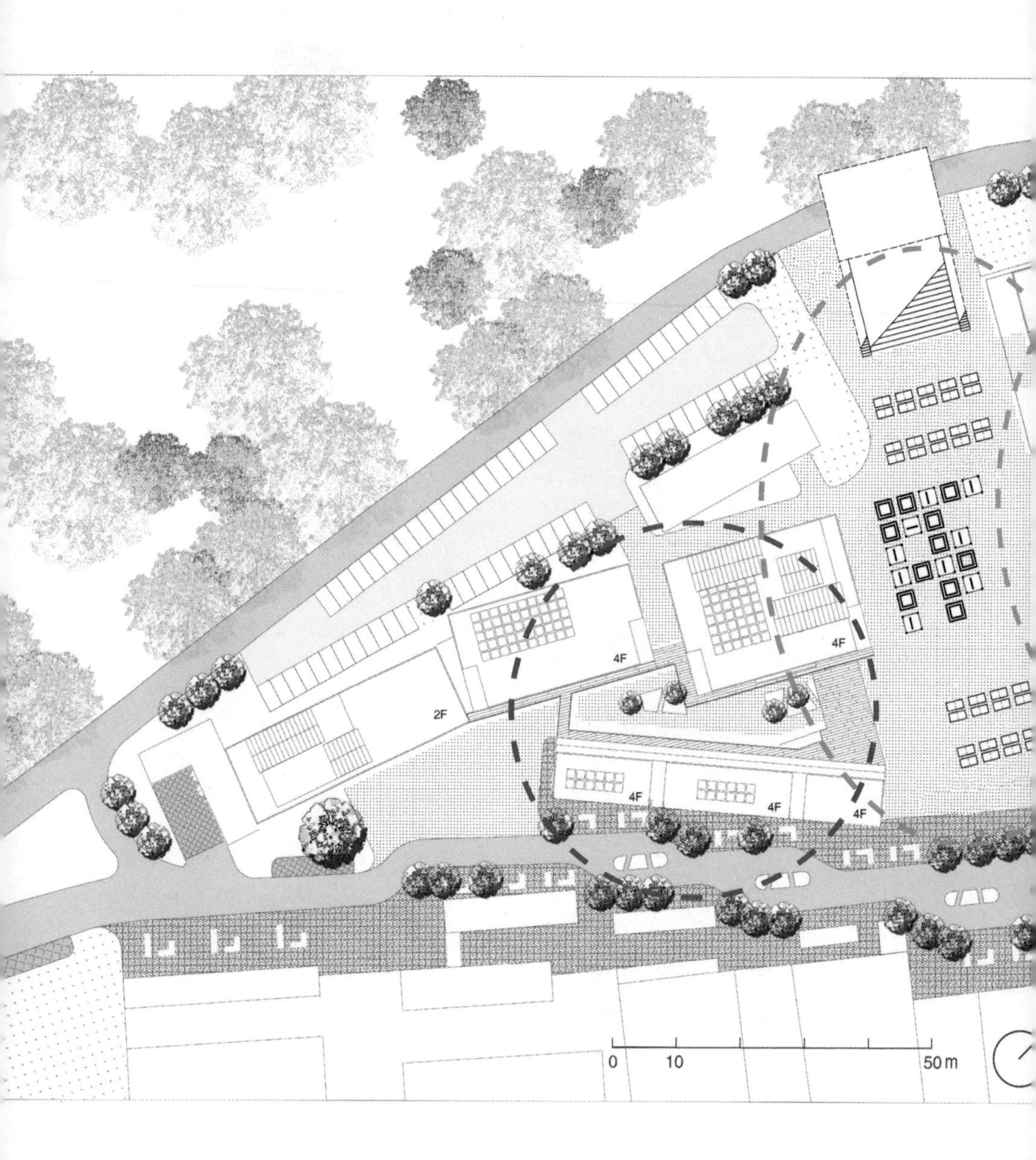

— — 第一步　　First Step

— — 第二步　　Second Step

— — 第三步　　Third Step

改造时序图

Renewal sequence diagram

ECO-CREATIVE

创意集市透视图
Creative market perspective drawing

ECO-CREATIVE

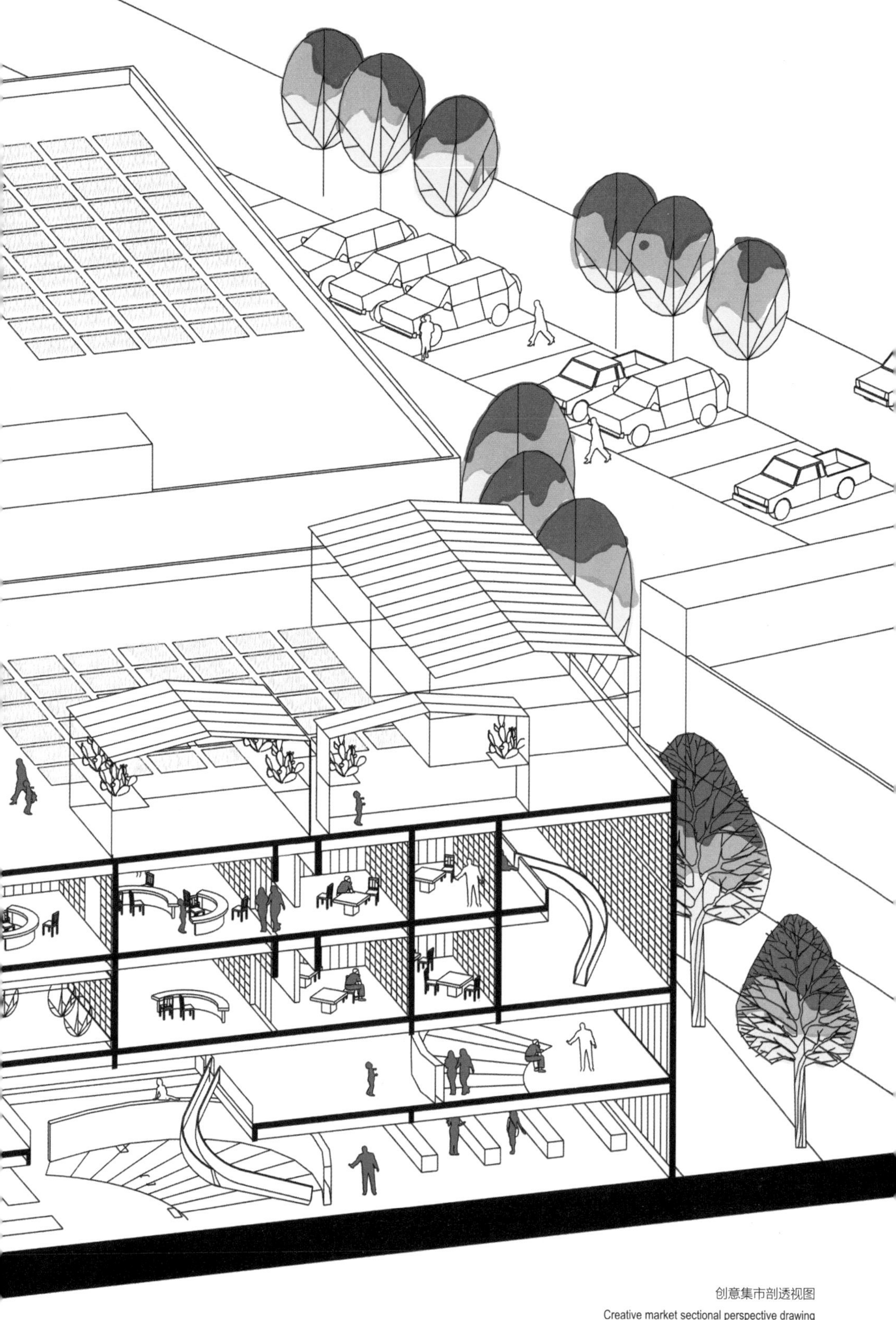

创意集市剖透视图

Creative market sectional perspective drawing

CREATIVE MARKET
FESTIVAL of PLAY

创意集市透视图

Creative market perspetive drawing

100

创意集市透视图

Creative market perspective

现状　Existing Situation

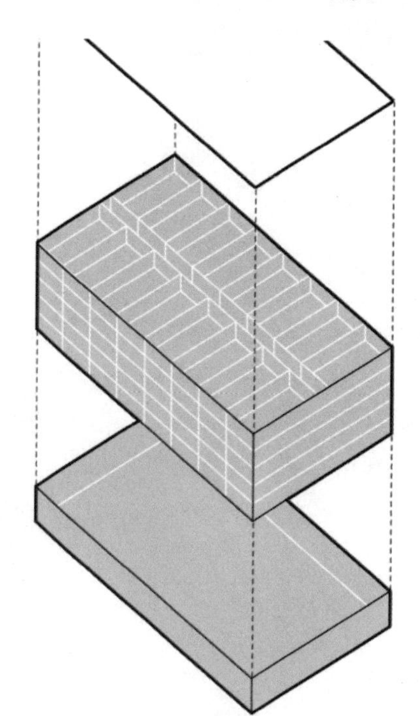

现状 Existing Situation

可能性 Possible Adaption

住宅　Residential

共享工作空间　Co-Working

商业　Commercial

展览　Exhibition

生态毂　Eco-Hub

设备　Facilities

改造策略图

Renewal strategies

建筑空间改造方式 Architecture Principles

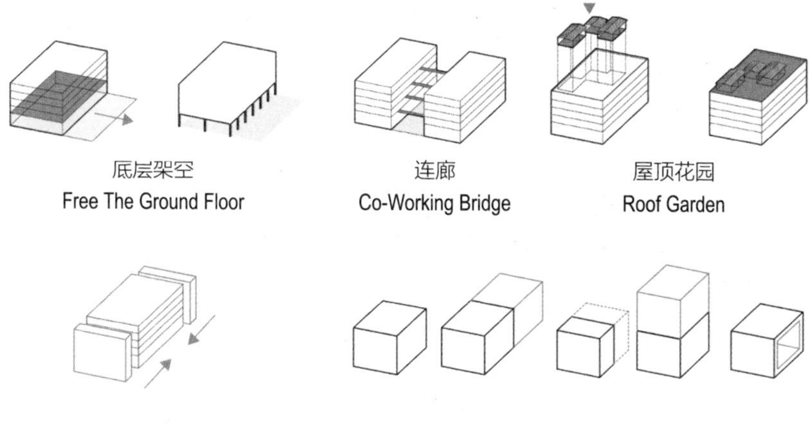

增加核心，开放空间
Add Core, Let The Workspace Open

工作区域的组合、堆叠、开放
Workspace Combined, Stacked Or Outdooe

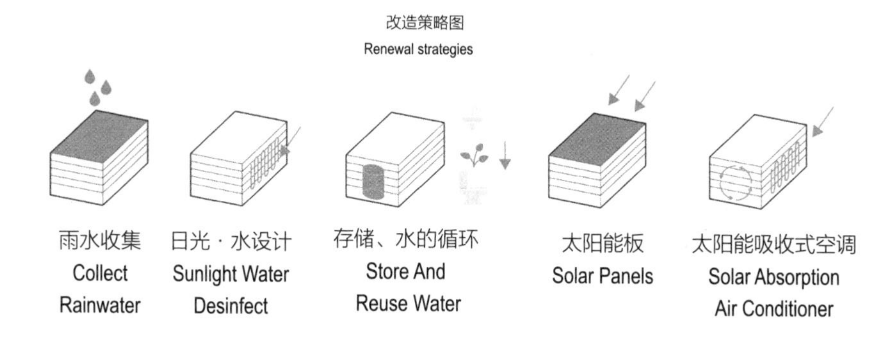

改造策略图
Renewal strategies

可持续性原则 Sustainability Principles

工厂| 创意办公室

现代工厂不适合在这个区域发展。 我们建议将其用作创意办公楼。通过对建筑物结构的微小和低成本修改，使其可以承载诸如初创公司办公室、共享工作空间、制造商车间设施和小型制造商等功能。 这能够为可持续发展领域的创造性思维营造一个丰富的环境。

Factories | Creative Office

The current use as factories is not suitable for this area anymore. We suggest the use as creative office buildings. This means with small and low-cost modifications to the structure of the buildings they can host functions like offices for start-ups, co-working spaces, makers spaces workshop facilities and small prototyping manufactures. This should create a bustling environment for creative minds in the field of sustainability.

现状 Existing Situation

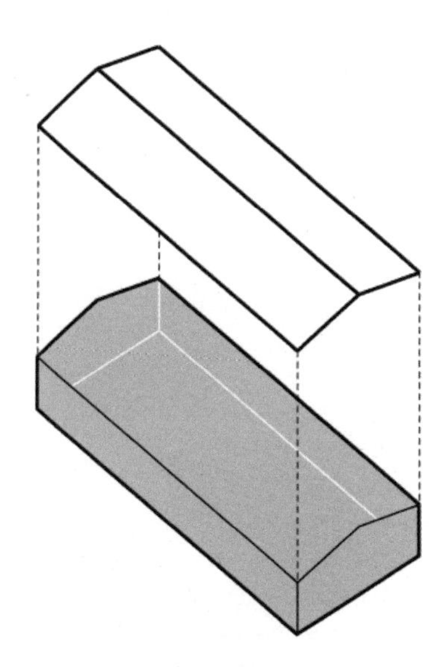

现状 Existing Situation

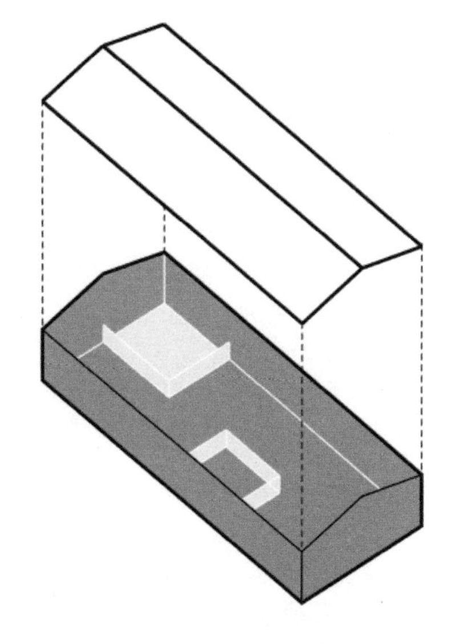

可能性 Possible Adaption

■	住宅	Residential
■	共享工作空间	Co-Working
■	商业	Commercial
■	展览	Exhibition
■	生态毂	Eco-Hub
■	设备	Facilities

改造策略图
Renewal strategies

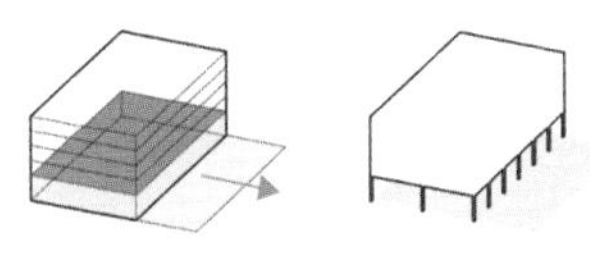

底层架空

Free The Ground Floor

建筑空间改造方式 Architecture Principles

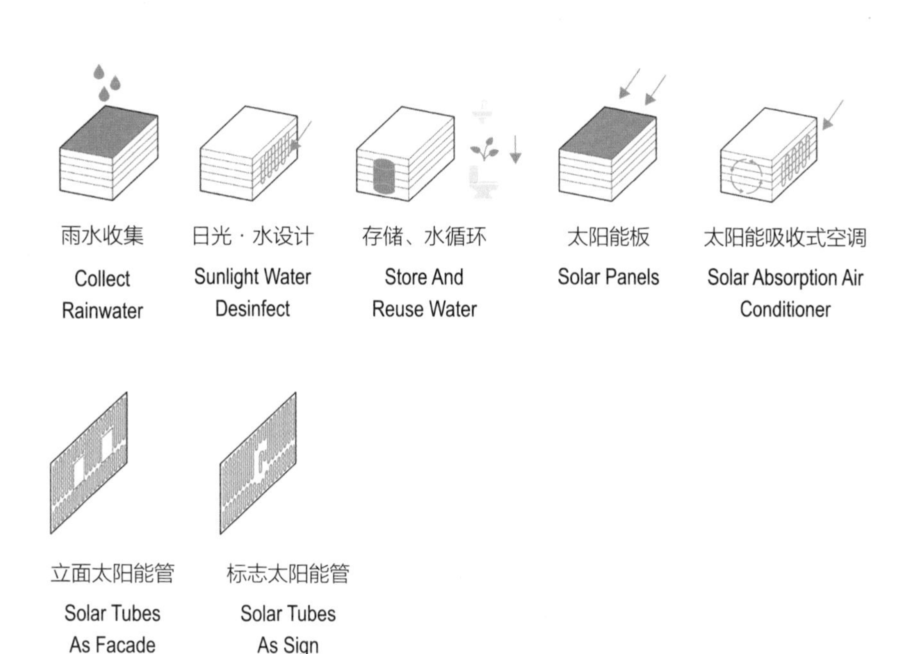

可持续性原则 Sustainability Principles

仓库|生态中心

唯一一幢这种类型的建筑物位于村庄的入口处。作为开发位置的完美起点，这个地方应该是创意和建立社区的交汇点。在这里，他们开始致力于创意工作、互相激励、共同庆祝努力的成果。

Warehouse | Eco-Hub

There is only one building of this type right at the entrance of the village. The location is the perfect start point for this development. This place should be the meeting point for the creative and start-up community. Here they start working on ideas, inspire each other and celebrate together.

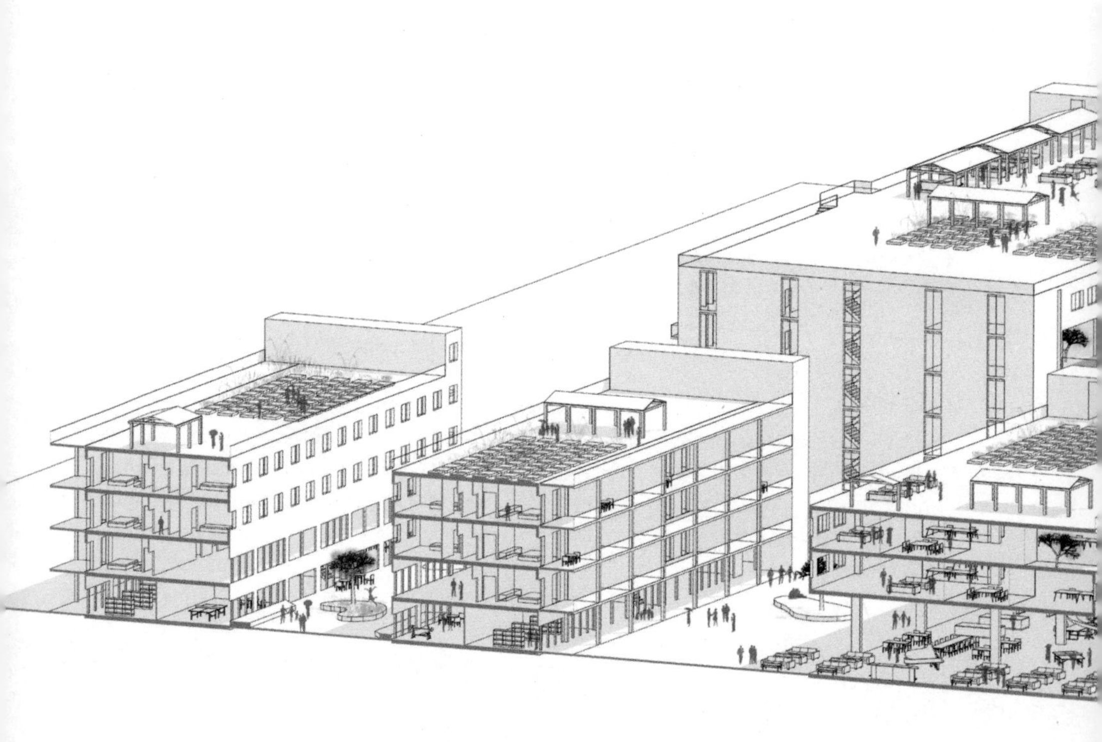

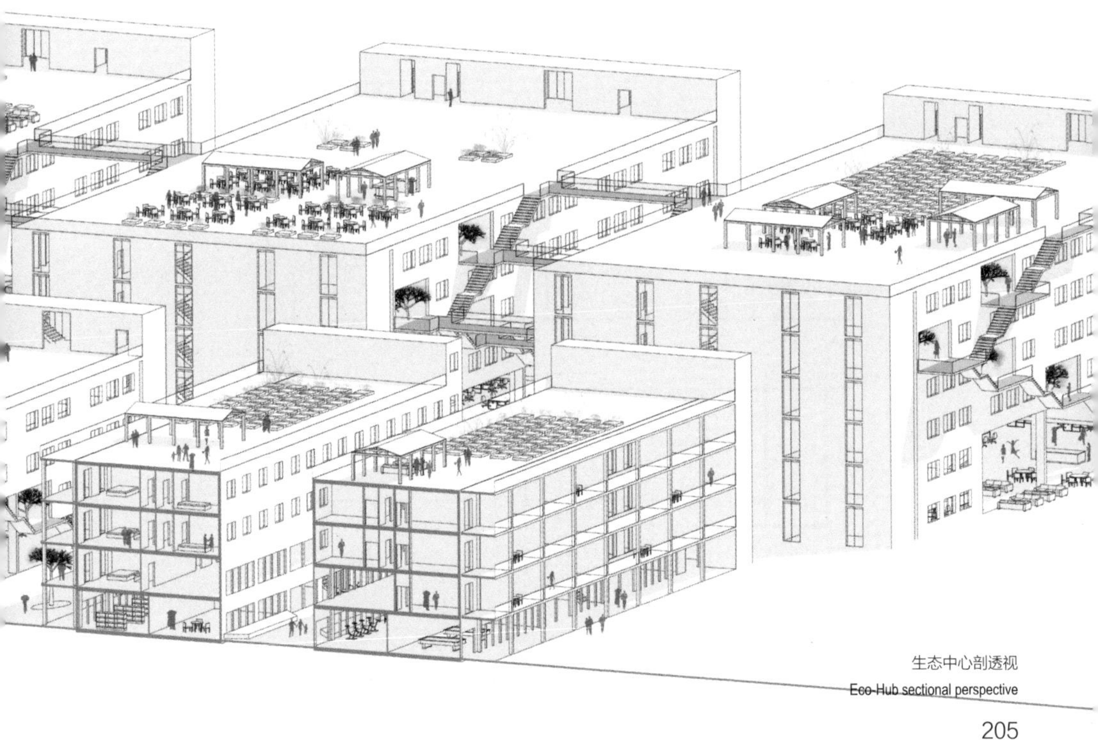

生态中心剖透视

Eco-Hub sectional perspective

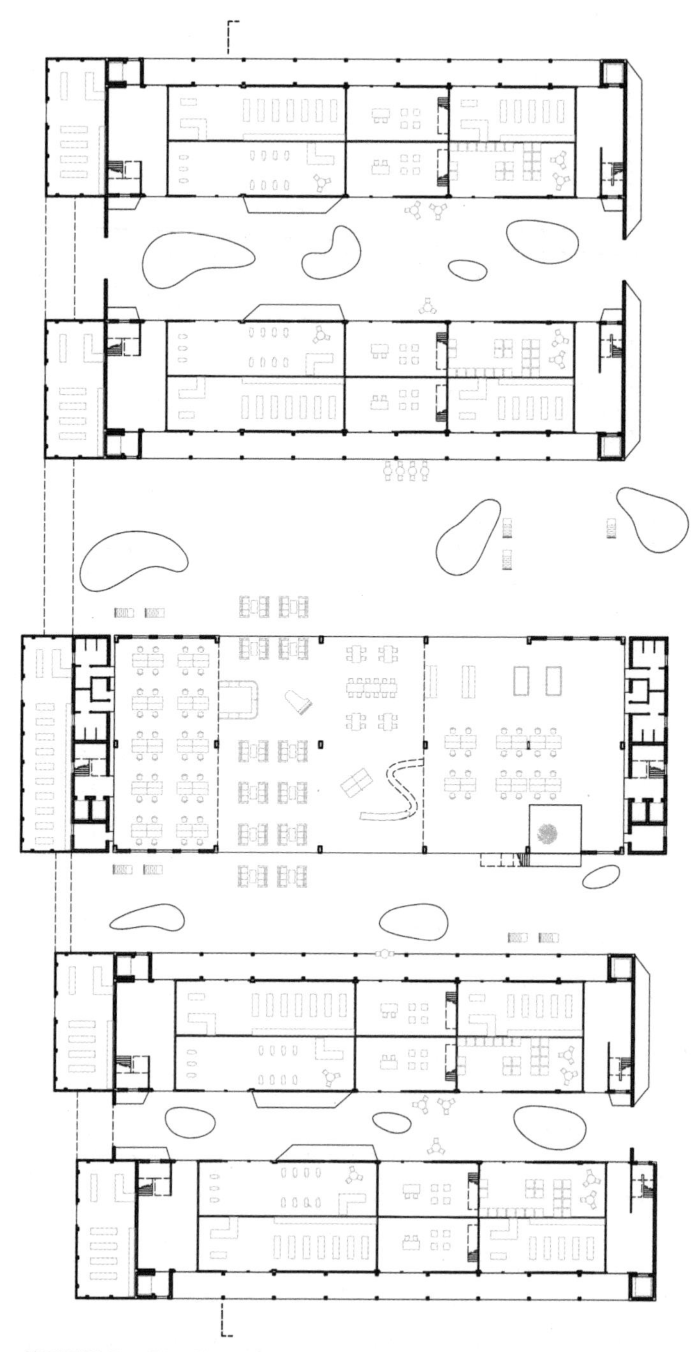

首层平面 First Floor Plan

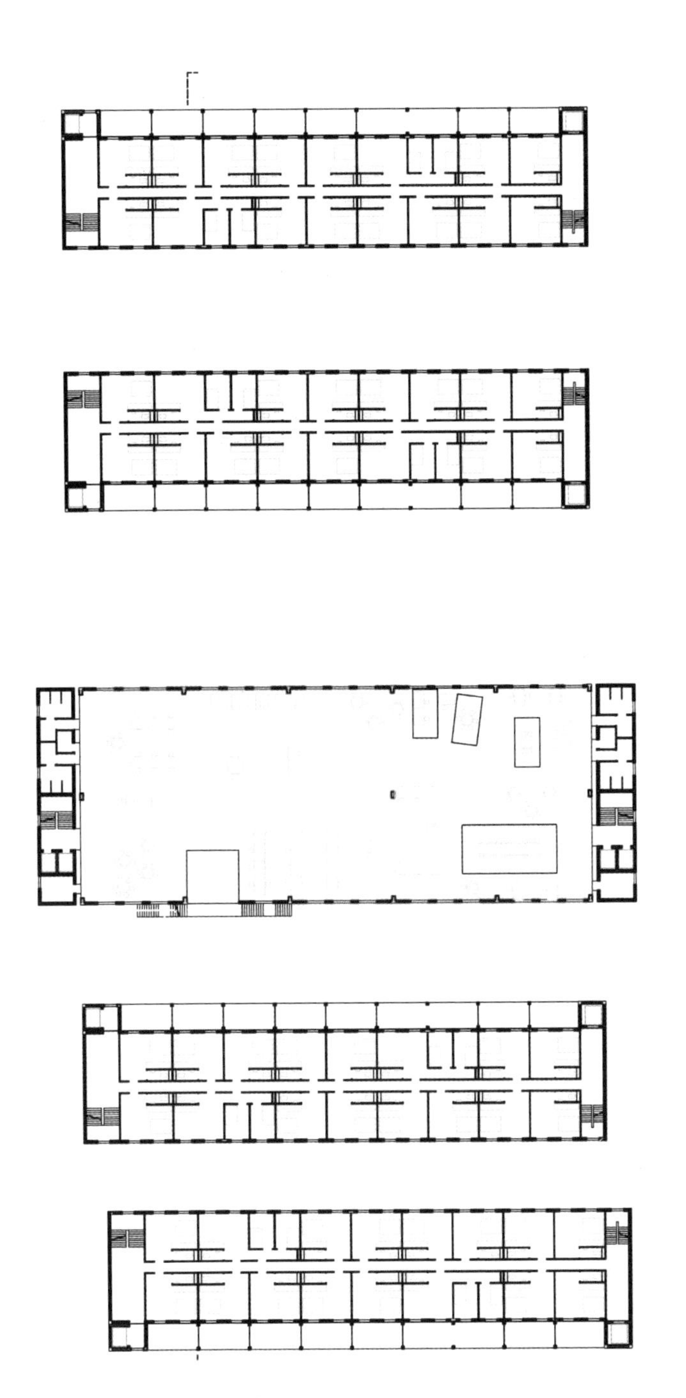

标准层平面　Regular Floor Plan

5.3 运动中心

SPORT CENTER

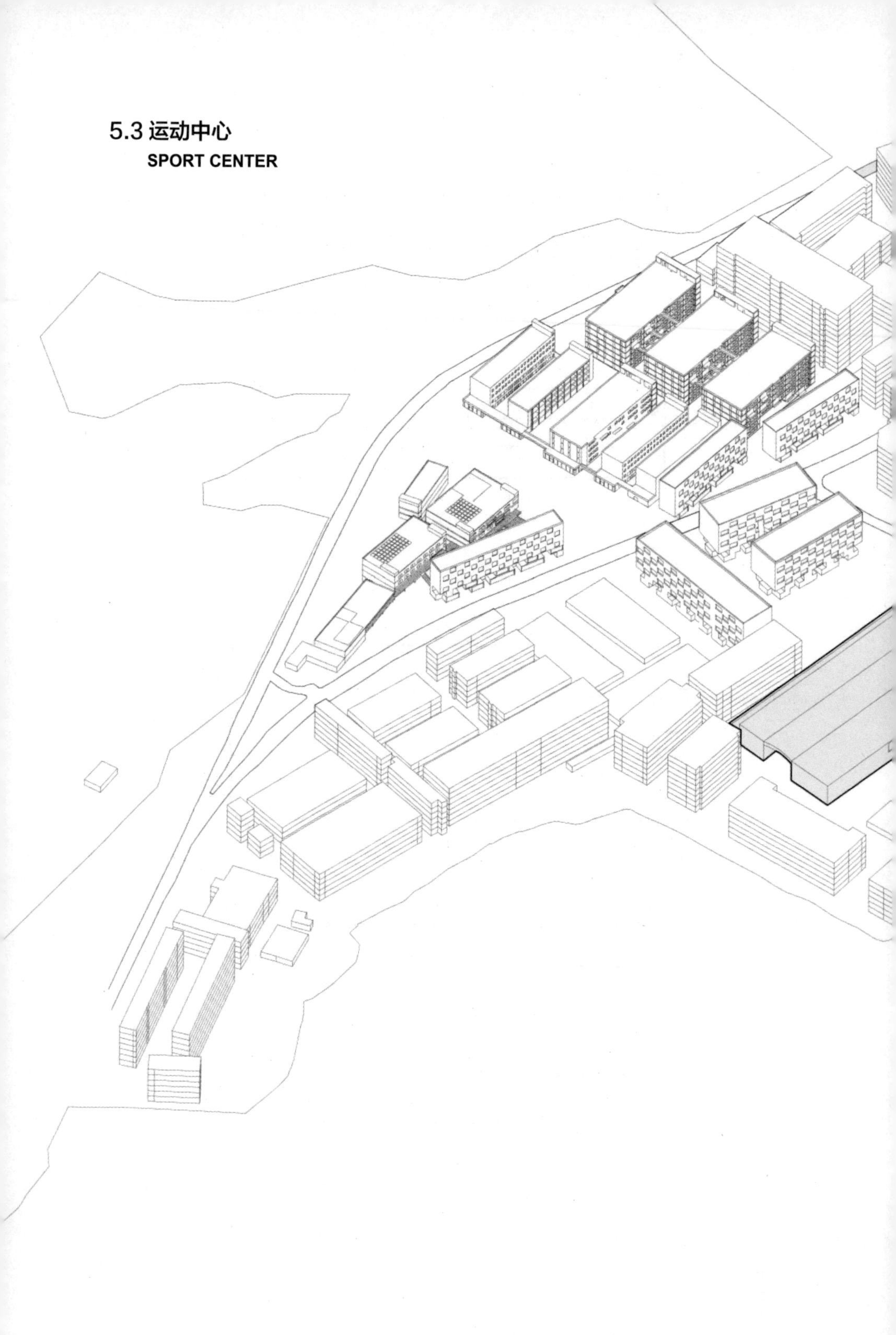

运动中心轴测图
Sport Center anaxometric drawing

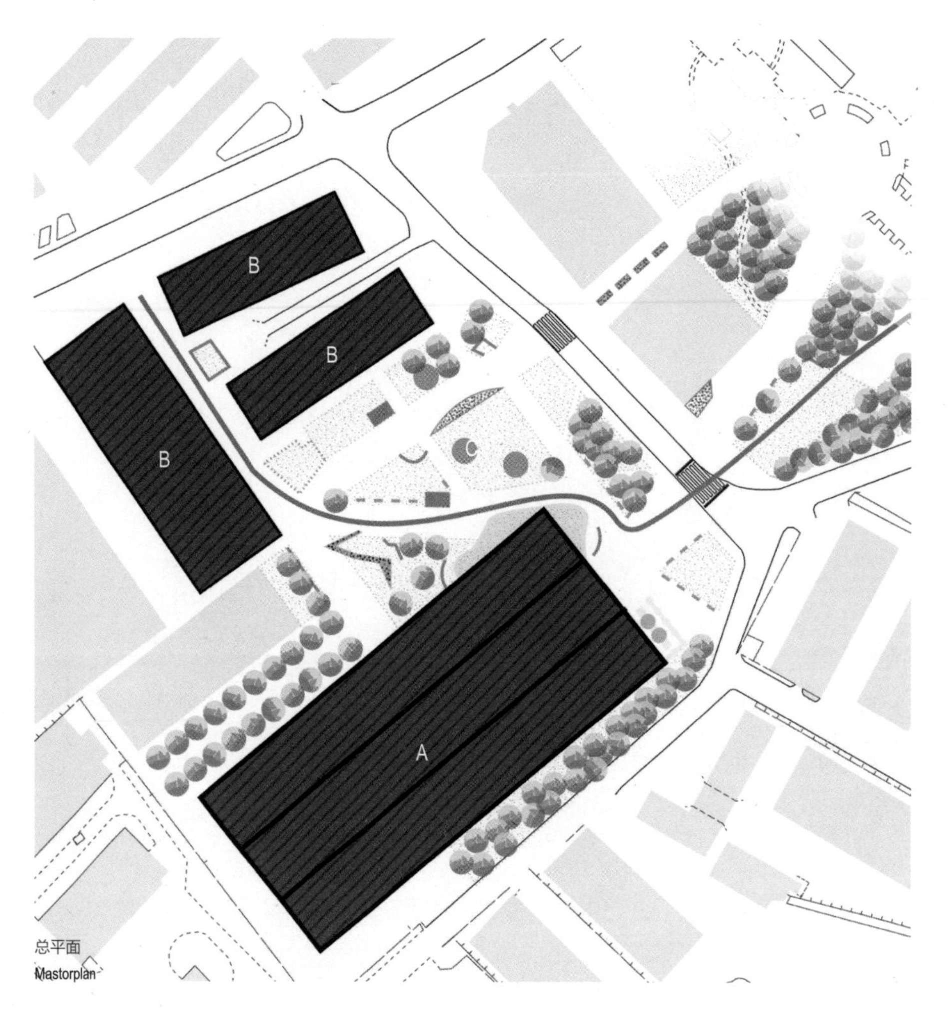

总平面
Mastorplan

A 体育中心 Sport Center

B 宿舍 Dormitory

C 公园体育活动 Park For Sport Activity

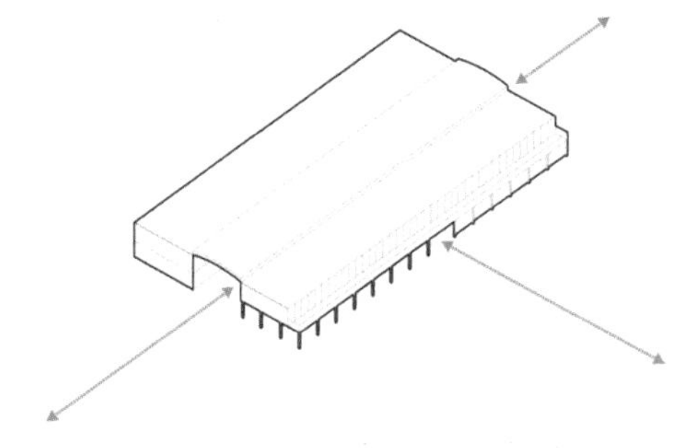

1.轴线 Axis

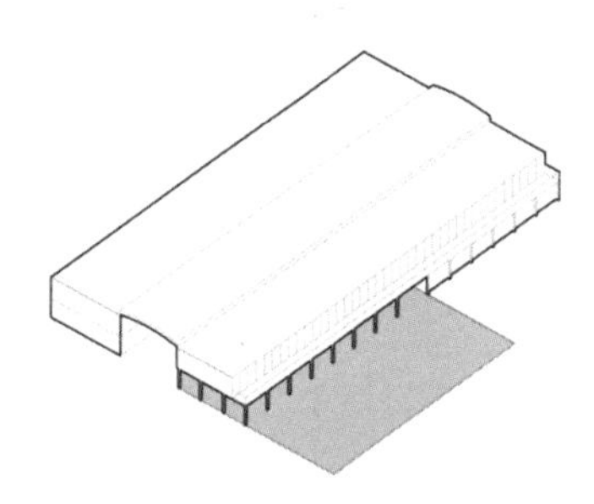

2. 开放式城市空间 Free Urban Space

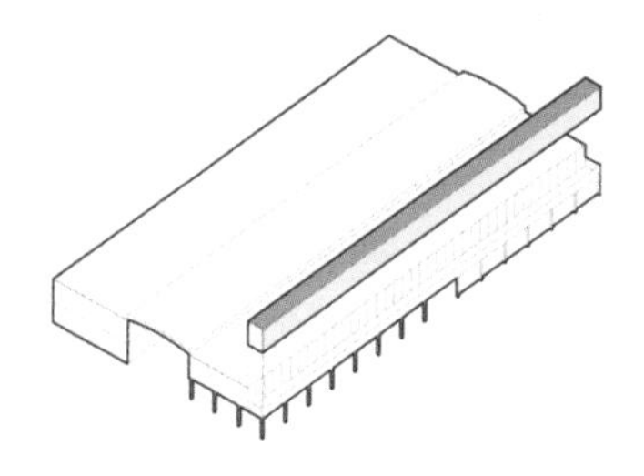

3. 视线延伸 Visual Connection To Square

4. 功能 Functions

改造策略图
Renewal strategies

运动中心改造概念

体育中心作为一个重要的、占主导地位的建筑，对周围的环境产生很大的影响。在我们的愿景中，它可以在未来作为不同体育活动的场所：滑冰场、羽毛球馆、篮球馆等。这一优点是没有任何结构变化，就可以为运动提供开阔空间。在设计中，我们对建筑首层进行了微调，在空间上实现与前方的广场更好地交流。

Concept of the refubrishment of the sport center

The sport center is just from its sizen an important, dominant building, that has a great influence on the surroundings.In our vision it can serve in the future as the spot of the different sport activities: from skating to badminton and basketball. The advantages of the building is that without any structural changes it can provide a space for sports, which requires a greater open space.In the intervention we would offer smaller changes in the ground level to create a better communication with the square in front.

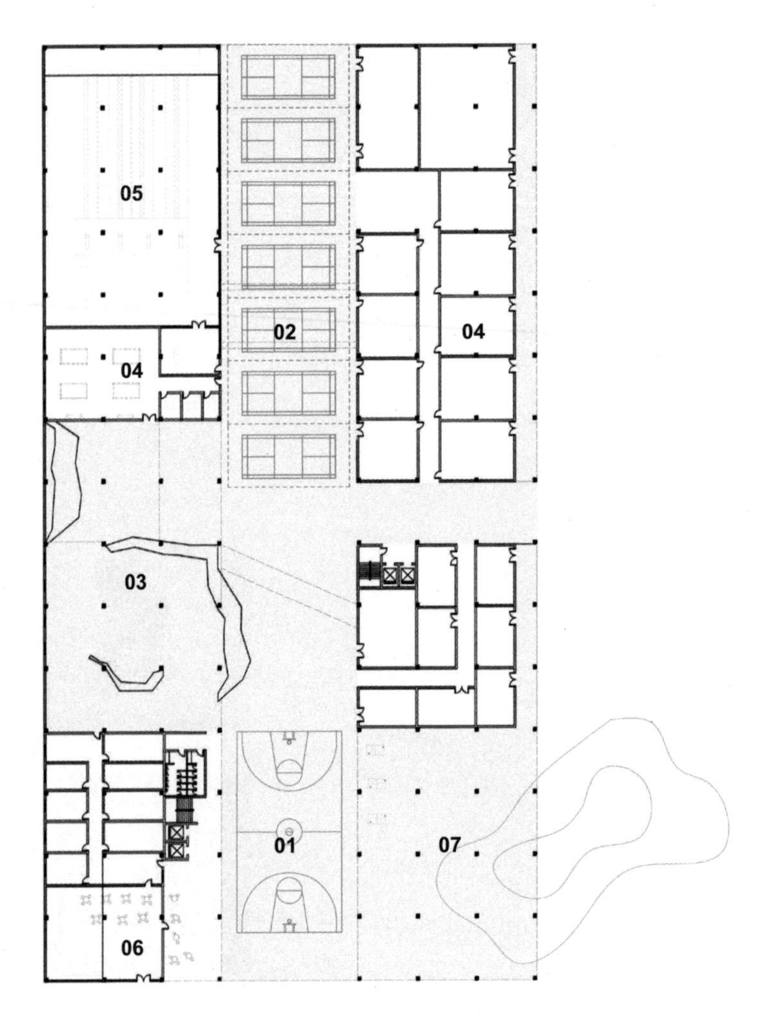

首层平面　　First Floor Plan

功能布局 Functional Layout

01	篮球场地	Basketball Field
02	羽毛球场地	Badminton Court
03	攀岩	Wall - Climbing
04	商店	Shops
05	仓储	Storage
06	餐厅	Restaurant
07	滑板	Skate Boarding

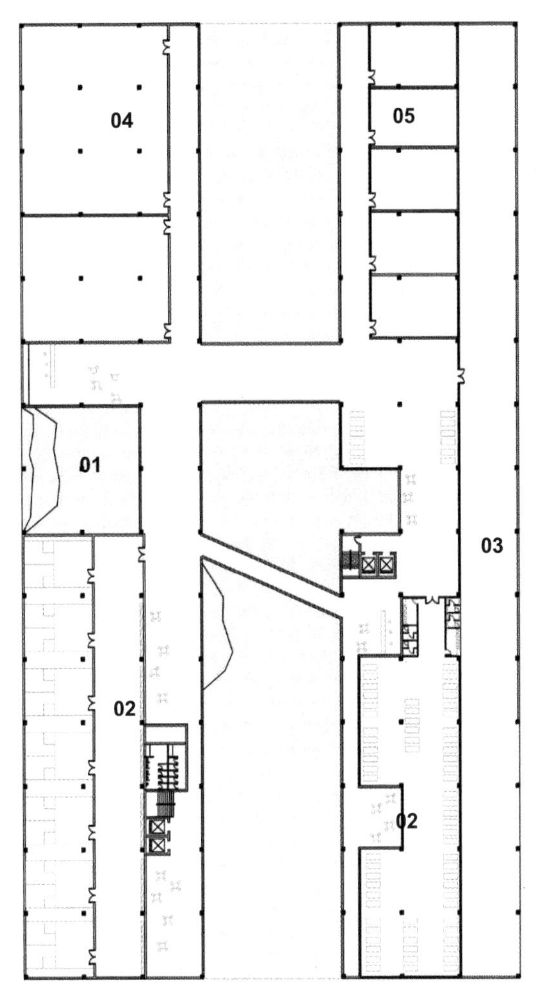

二层平面　　Second Floor Plan

功能布局 Functional Layout

01	攀岩	Wall - Climbing
02	体育馆	Gym
03	露台	Terrace
04	激光标签	Laser Tag
05	行政	Administration

运动中心透视图

Sport center perspective

5.4 公共建筑
PUBLIC BUILDINGS

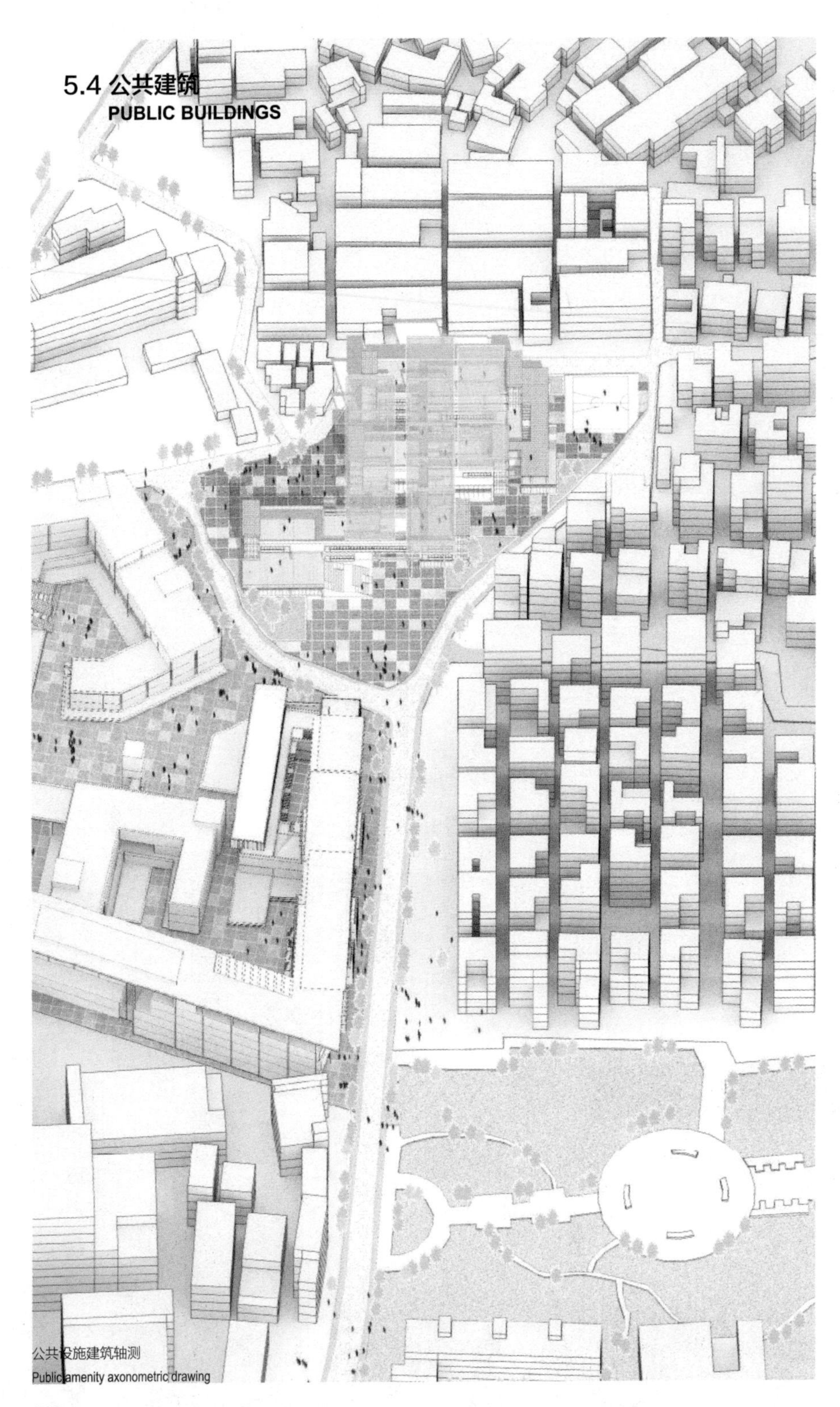

公共设施建筑轴测
Public amenity axonometric drawing

针对深圳典型的“城中村”现象，结合麻磡村作为深圳市水源保护地，急需进行产业调整的现状设置，希望通过对该村功能与空间形态的更新，实现麻磡村的再生。

在居住片区结合附近废弃的学校和空地，形成村子的公共活动中心，并结合商业、培训、运动等功能的植入。通过次级街区的围合形成一系列运动公园，并辐射至周边的居住地块，形成具有中心性的空间核心。同时在宏观层面，以组团方式，构建能够彼此衔接的公共空间网络，将原来线性的村子结构转变为以主街为轴，以公共空间主干为环的复合结构，使其具备更强的复杂性和整体关联性，为未来麻磡村的空间发展制定下基本的逻辑框架。在微观层面做到既存建筑的最大化利用。这种利益上的最大化不仅体现为对用地建筑密度的提升，更在于对近地面层空间价值的开发和对建筑混合利用的功能组织。

In view of the typical “village in the city” phenomenon in Shenzhen, combined with paralysis as the water source protection site in Shenzhen, it is urgent to set the status quo of industrial adjustment. It is hoped that the regeneration of the village will be realized through the renewal of the function and spatial form of the village.

In the residential area, combined with abandoned schools and open spaces, the villages' public activity center is formed, and combined with the functions of business, training, sports and other functions.A series of sports parks are formed through the enclosure of the secondary blocks, and radiated to the surrounding residence, with the plot being formed a central core of space. At the same time, at the macro level, we examined the distribution of the public space around the main street, and built a public space network that can be connected to each other in a group manner, transforming the original linear village structure into a new structure with the main street as the axis and public space as trunk. As a composite structure of the ring, it has greater complexity and overall relevance, and lays a basic logical framework for the future development of paralysis. At the micro level, we propose the principle of high-intensity land use through the study of the value of land in Shenzhen, so as to maximize the use of existing buildings. The maximization of this kind of interest is not only reflected in the improvement of the density of land used for construction, but also in the development of the spatial value of the near-surface layer and the functional organization of the mixed use of buildings. To this end, we have established a three-dimensional trail system as an extension of the public space on the ground floor and extended to the interior of the existing building, which becomes a prerequisite for further reconstruction of the building space.

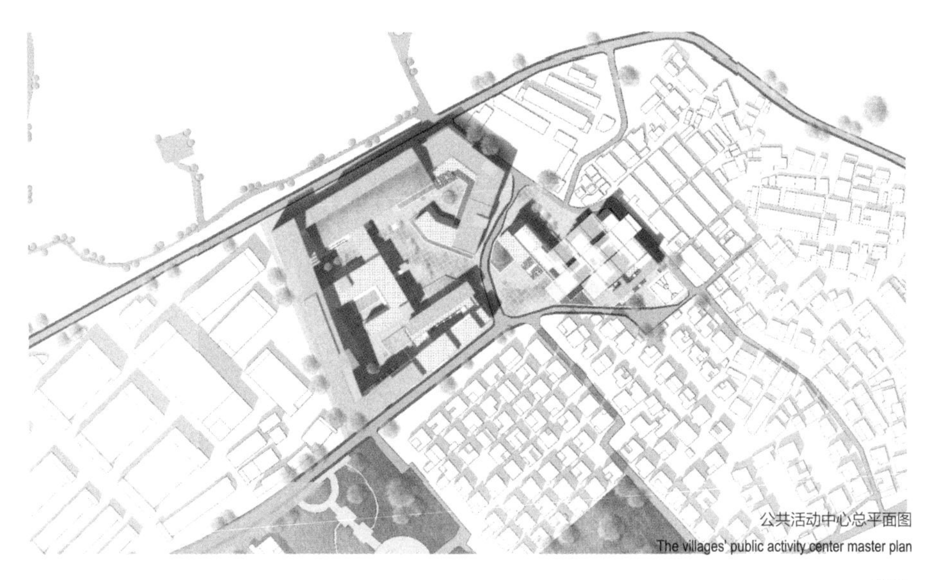

公共活动中心总平面图

The villages' public activity center master plan

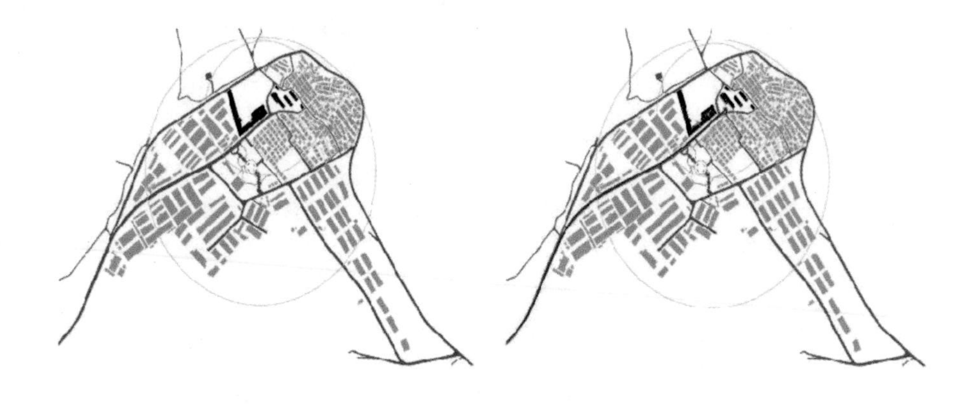

1 废弃的小学开放空间和分散式住宅

Village center, abandoned primary school open space & scattered residence

2 街道界面延伸，展开界面姿态

Continues the street section, open interface posture

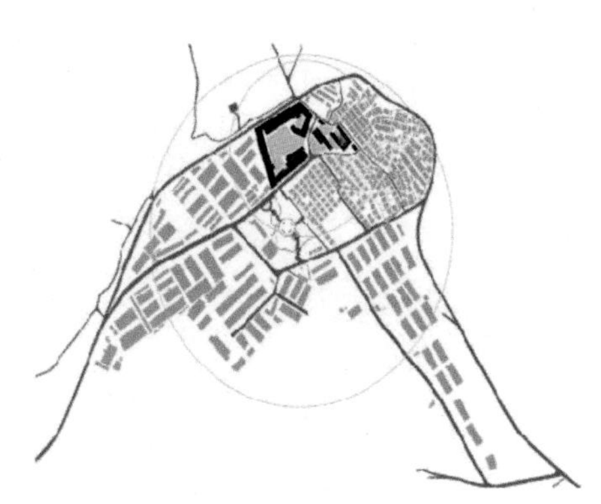

3 围合式界面形成公共庭院和中心剧院

Interface enclosure, forming a public space courtyard & center theater

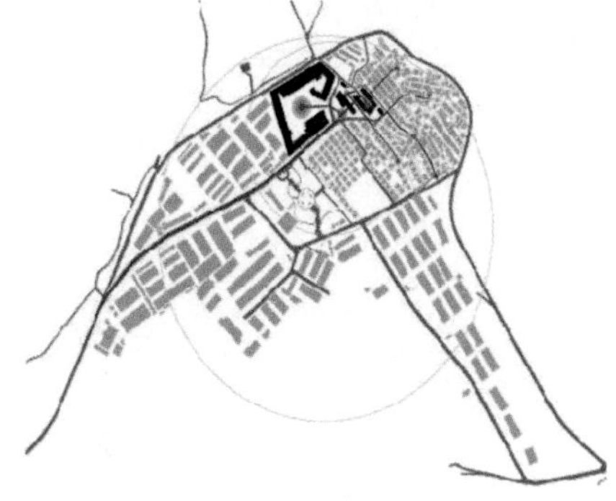

4 中心公共空间，辐射周围节点

Hinterland public space connected, radiating village nodes

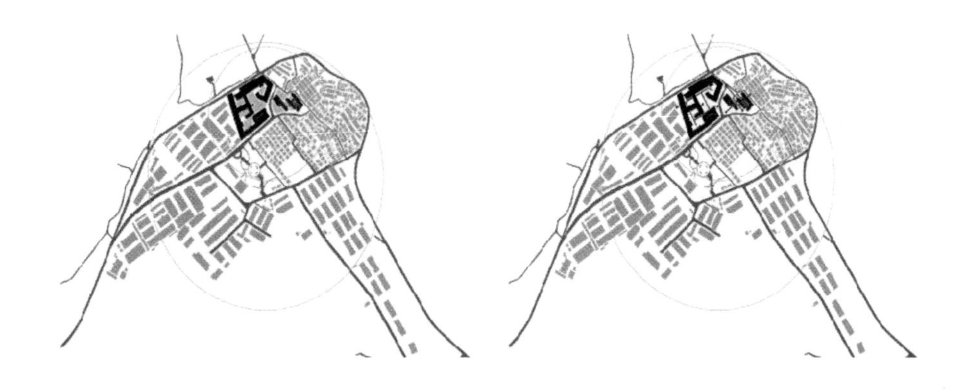

5 充实内部空间，丰富空间院落层次

Fill the internal space and enrich the courtyard level

6 路径贯穿，形成慢行系统

The internal path runs through and the two sites form a series walking system.

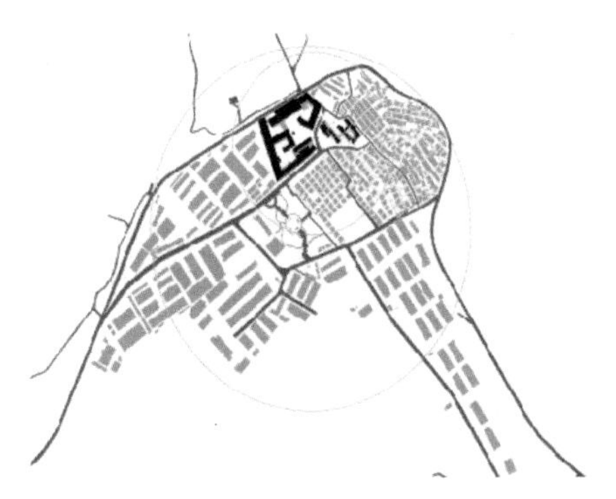

7 活动平台及廊道系统

Activity platform and gallery system

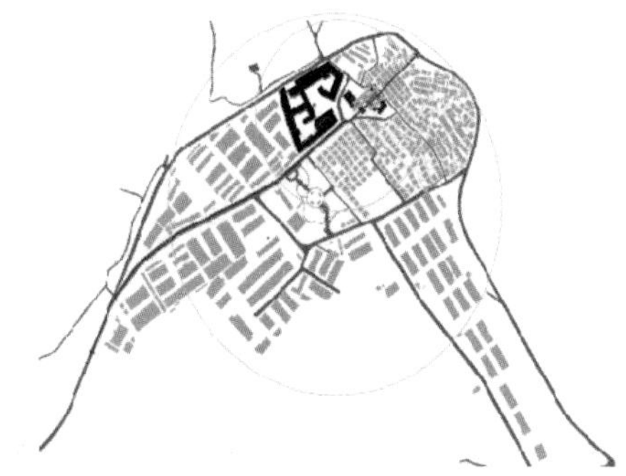

8 场地中心成为界面结构的延续

The center of the site has become a continuation of the main street structure

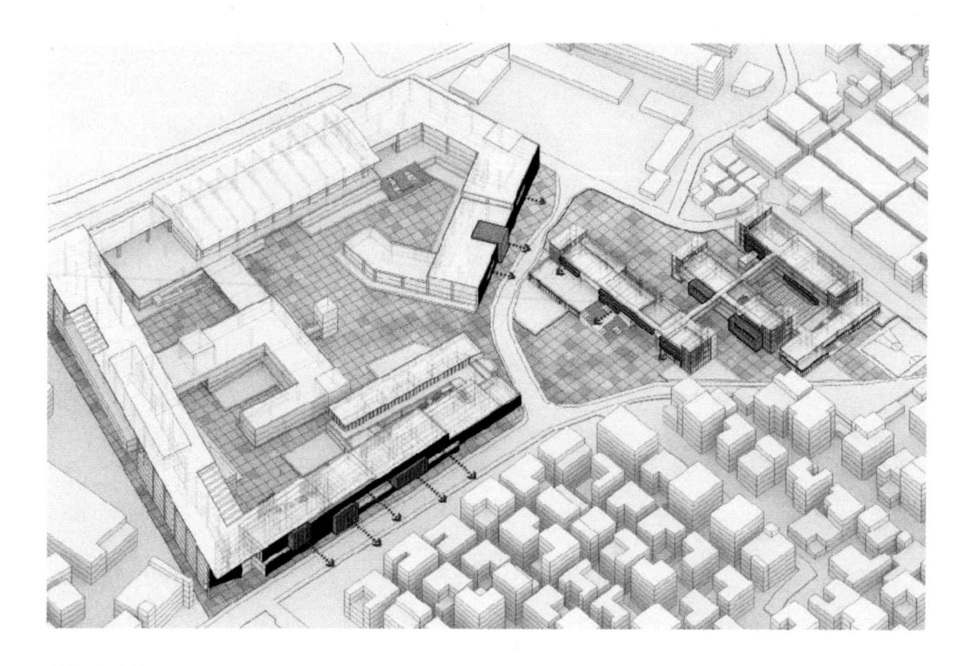

街道—界面关系
Stree-Interface Relationship

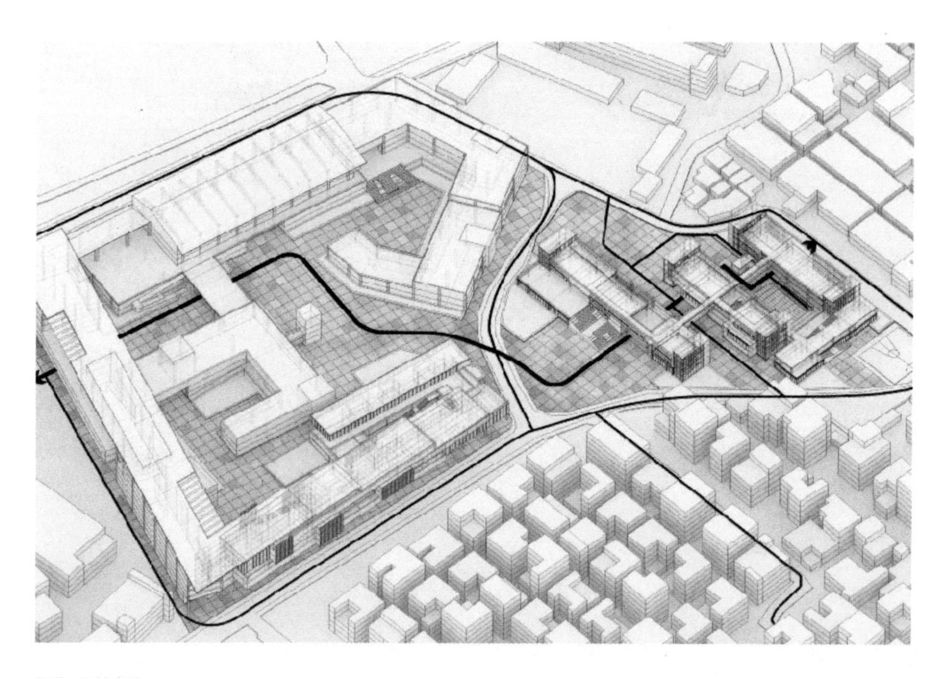

街道—场地失联
Street-Site Association

Free Travel
自由流线

Public Space System
公共空间系统

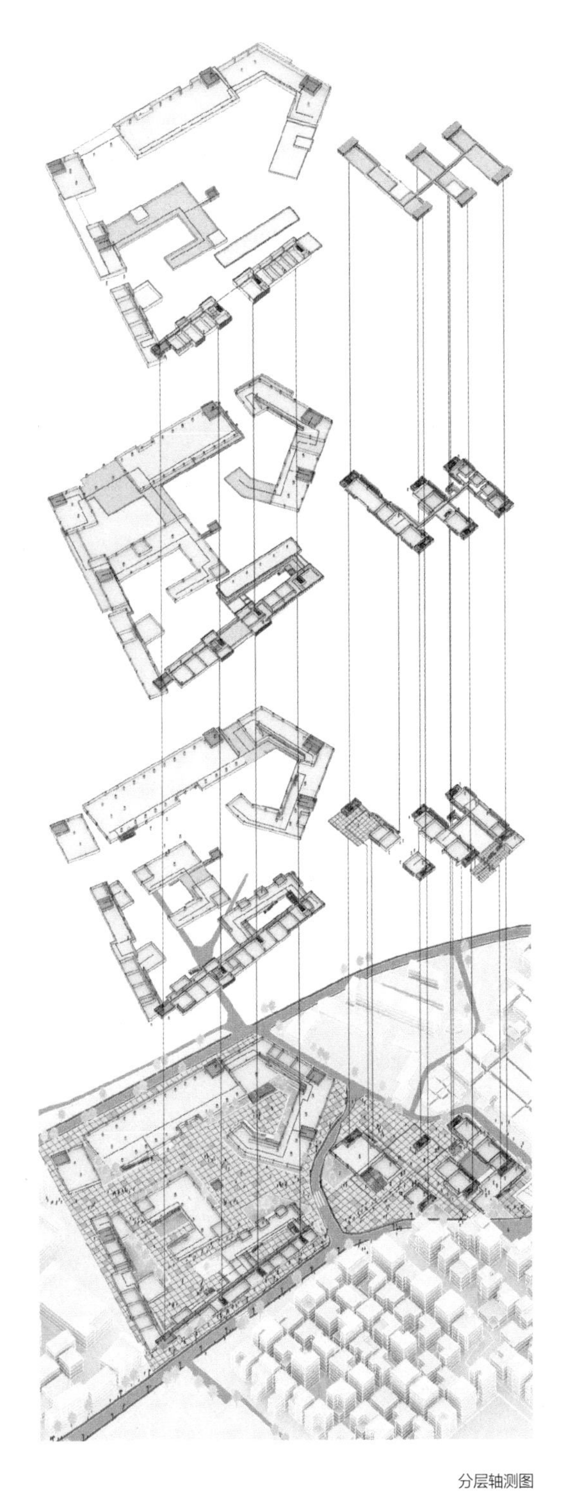

分层轴测图
Explosive axonometric

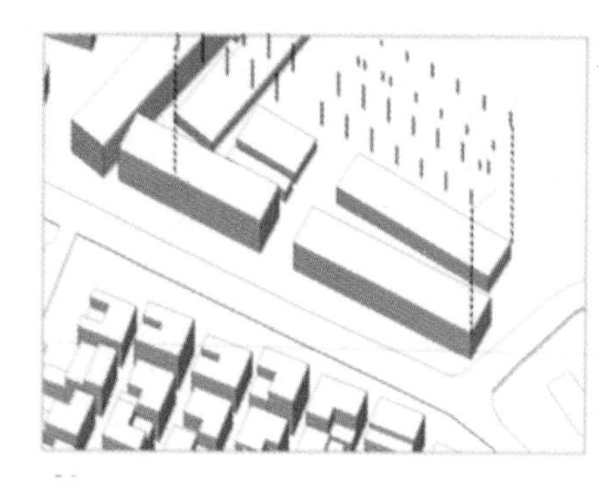

1 预留柱网

Reserved column network

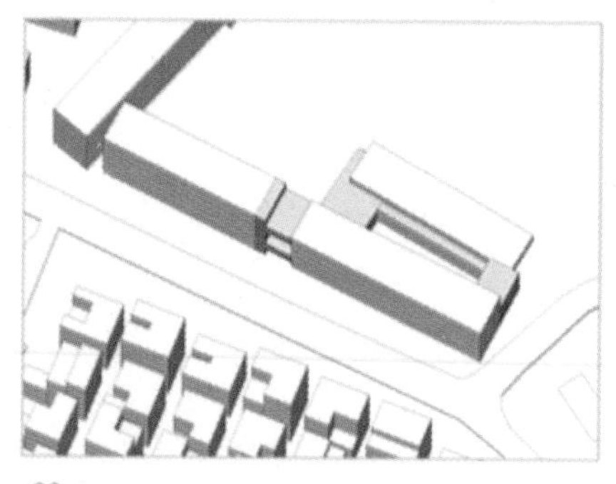

2 连接建筑，围合庭院

Connecting buildings, enclosing the courtyard

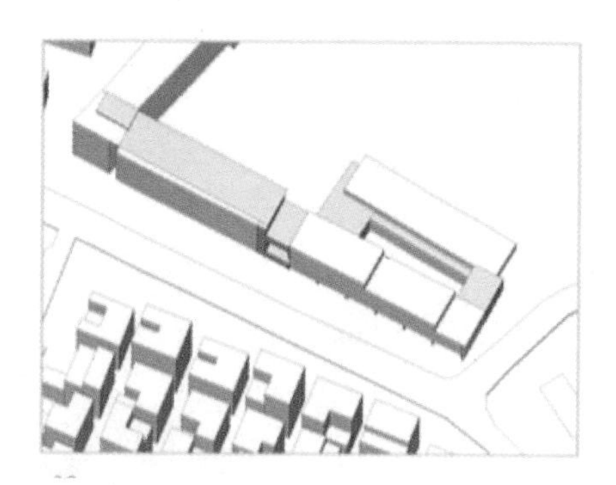

3 屋顶指向着腹地

The roof is facing the belly, pointing

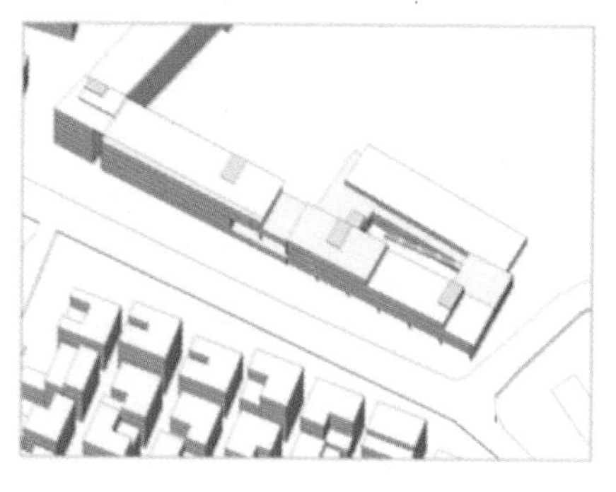

4 建立平台系统，竖向交通

Establish platform system, vertical traffic

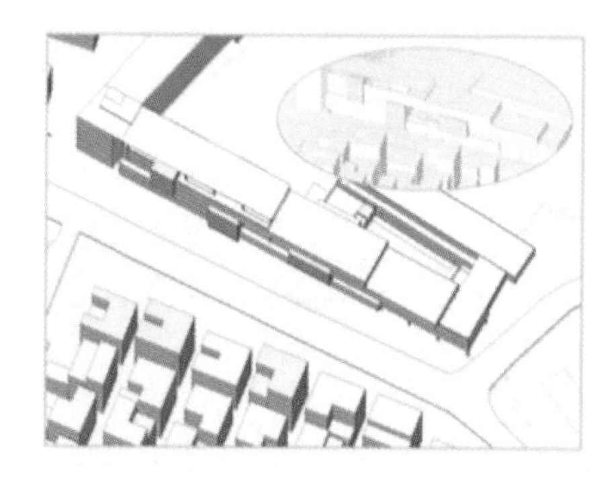

5 提取原型，使体量与立面有机融合

Extract the prototype to make the volume and the facade organic

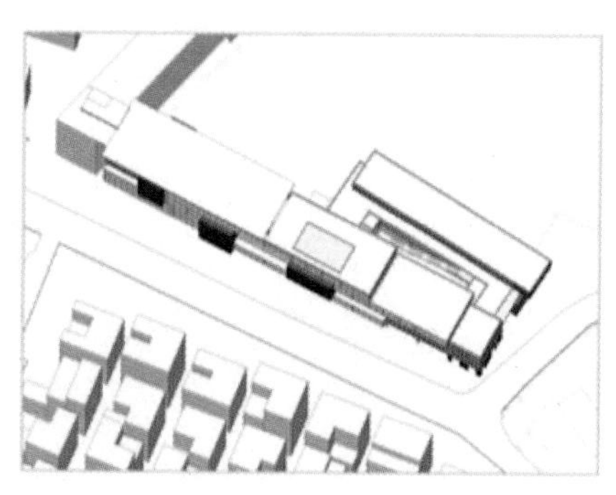

6 植入功能，绿色引入建筑内部

Implant function, green introduction inside the building

形体衍生

Shape Generation

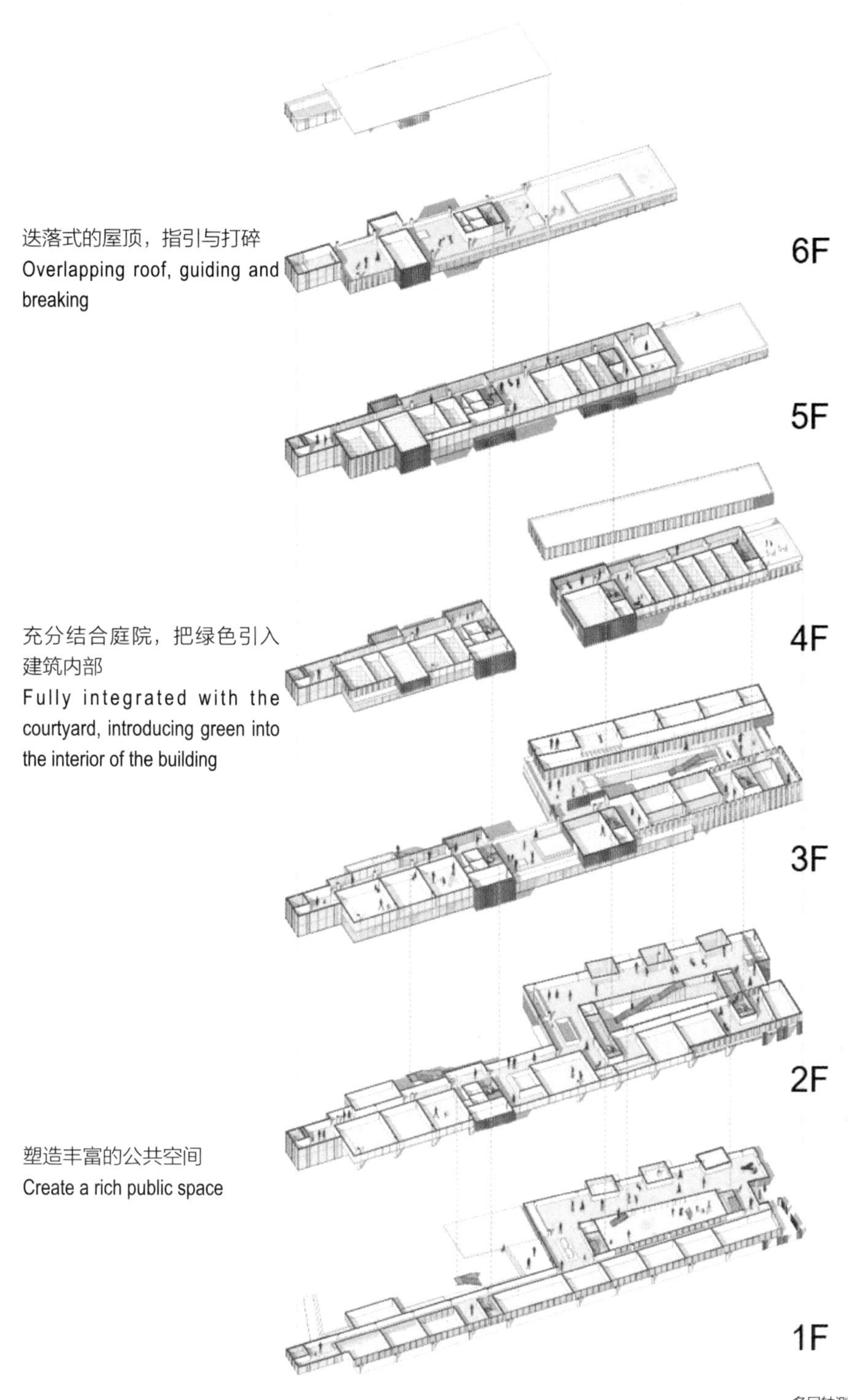

多层轴测图
Explosive axonometric

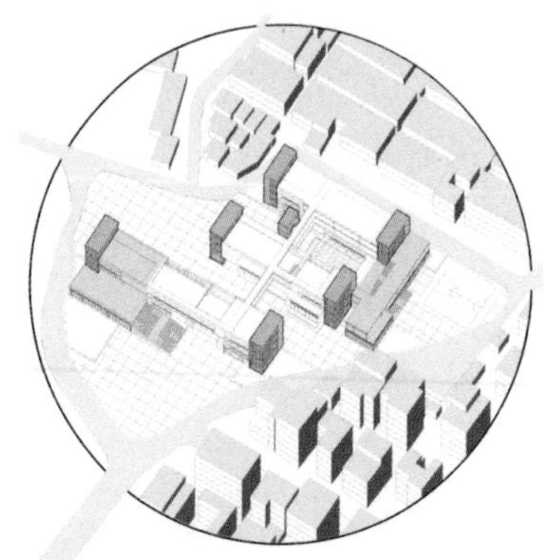

1 原建筑局部体量的架空与新建
Overhead and new construction of the local volume of the original building

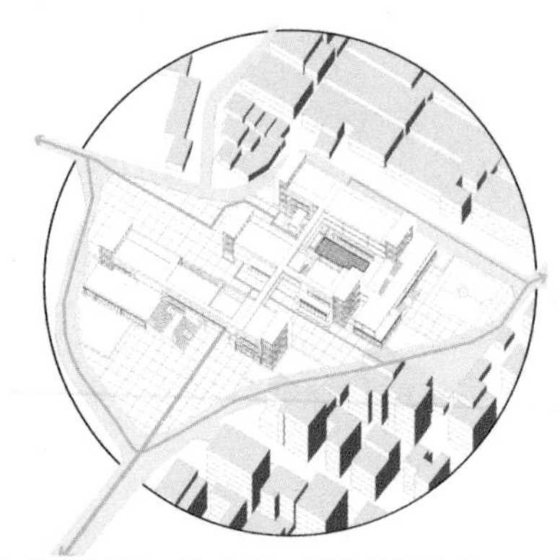

2 路径转折，为主街、村落次要出入口、北侧居住区人流提供聚集性公共空间
Path turns, the north side residential area provides a concentrated public space

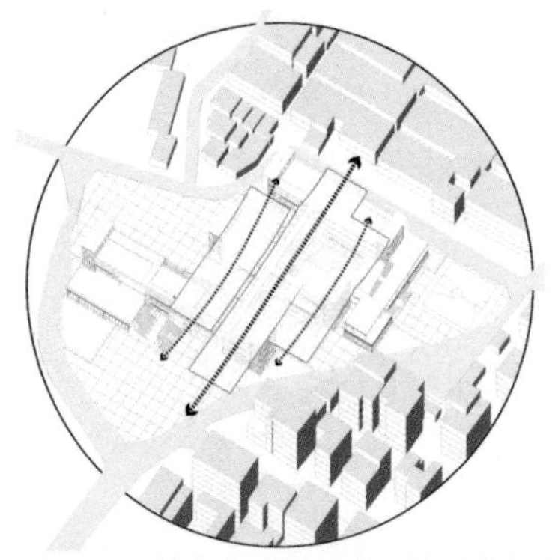

3 大屋顶延续主街空间秩序，从形态角度强调村落中心地位
The big roof continues the main street space order

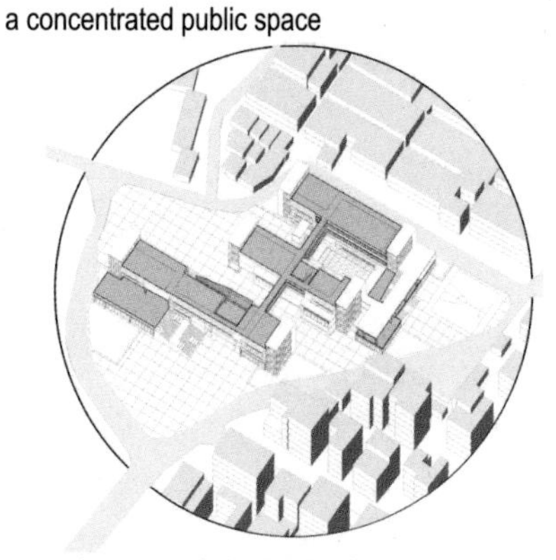

4 活动平台，连廊，外挂楼梯提供室外活动空间
Activity platform, gallery, external stairs provide outdoor space

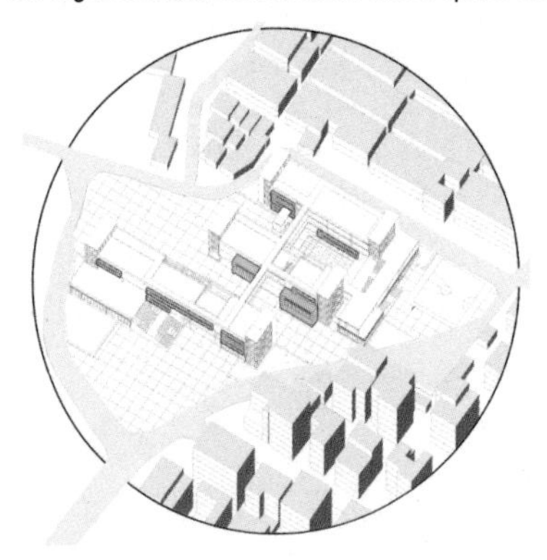

5 置入体量，打破原有单廊式空间组织，并提供较大空间满足室内公共活动
Put in the volume, break the original single-roof space organization, and provide more space to meet indoor public events

功能布局
Functional Replacement

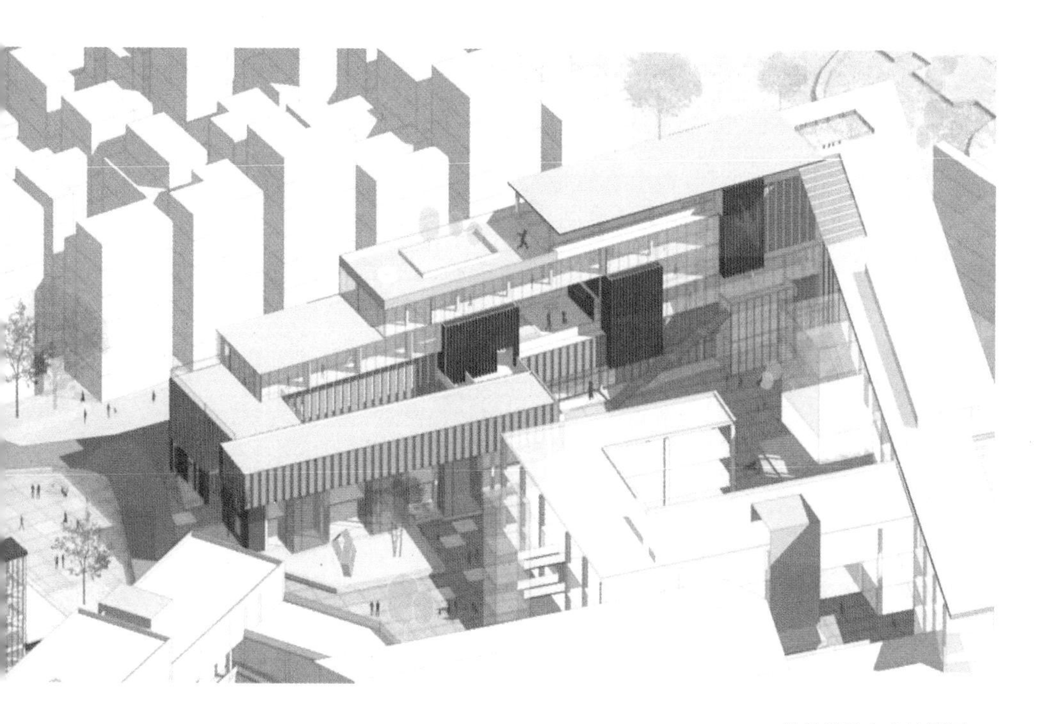

公共活动中心轴测图
The villages' public activity center

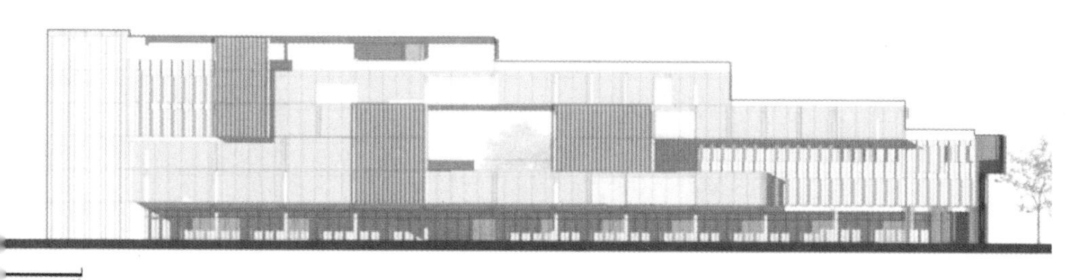

北立面 North Facade

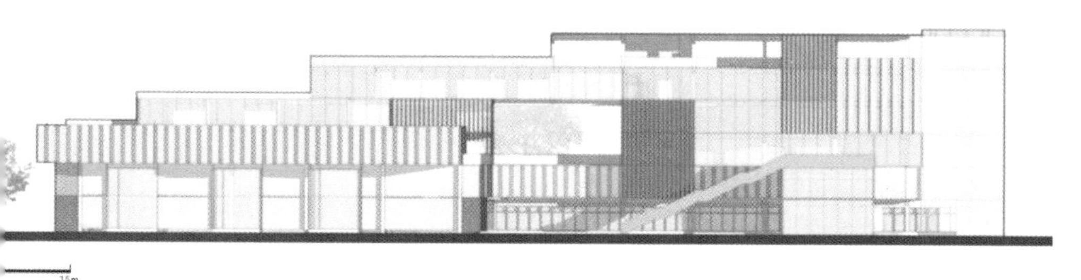

南立面 South Facade

位于城市水源地的麻磡村，兼顾着城市与自然衔接的使命，生态环境修复和公共空间体系紧密联系。现有麻磡村的建设并没有重视景观与环境的价值，综合整治将利用景观生态资源打造优质公共城市空间体系，营造活力社区，提升整村人居环境，进一步促进城中村整体升级。

Located at the urban water source, Makan village is response to connect the city with the nature. Ecological repair goes with the construction of public space of the village. The current construction of Makan village ignores the importance of ecological environment. An alternative method of renewal will be introduced, utilizing the ecological resources to promote the urban public system and creating vitality for communities, further improve the living environment.

6 麻磡村的改造实验——公共空间整治
城市活力的塑造

MAKAN RENOVATION STUDY-PUBLIC SPACE
URBAN DYNAMIC GENERATOR

6.1 界面激活

DEVELOP INTERFACE ACTIVATION

主街轴测图

Main street axonometric

butique

主街透视图 I
Main street perspective

主街透视图 II
Main street perspective

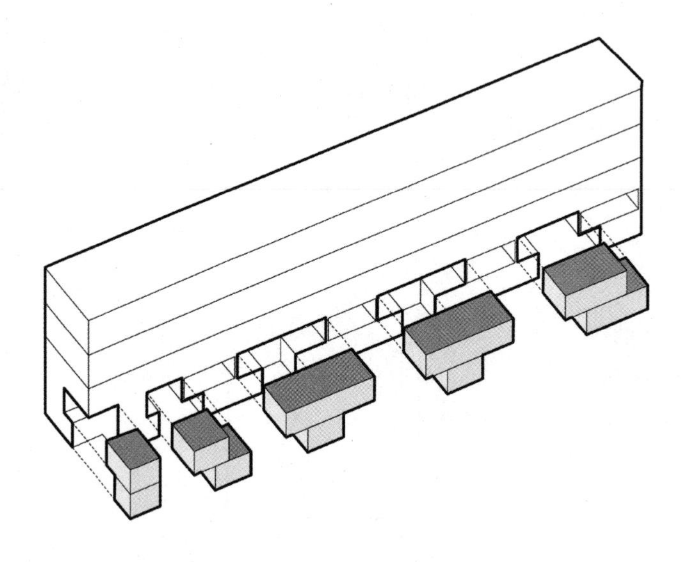

去除一楼部分体量
Removing Volumes in First Floor

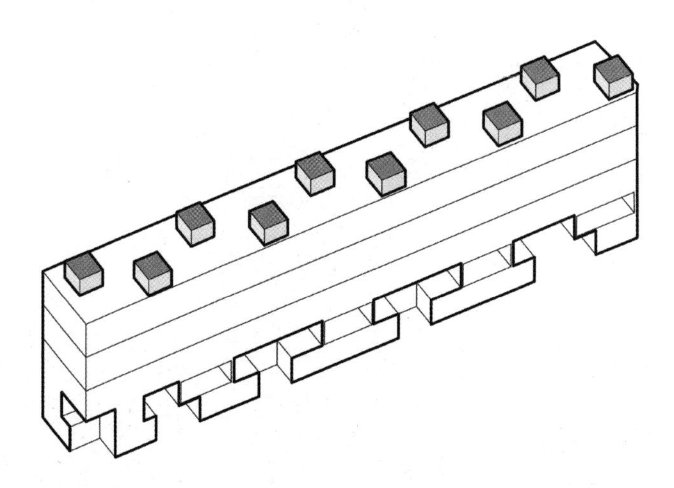

将移除的体量加到屋顶
Adding removed Volumes on Roof Top

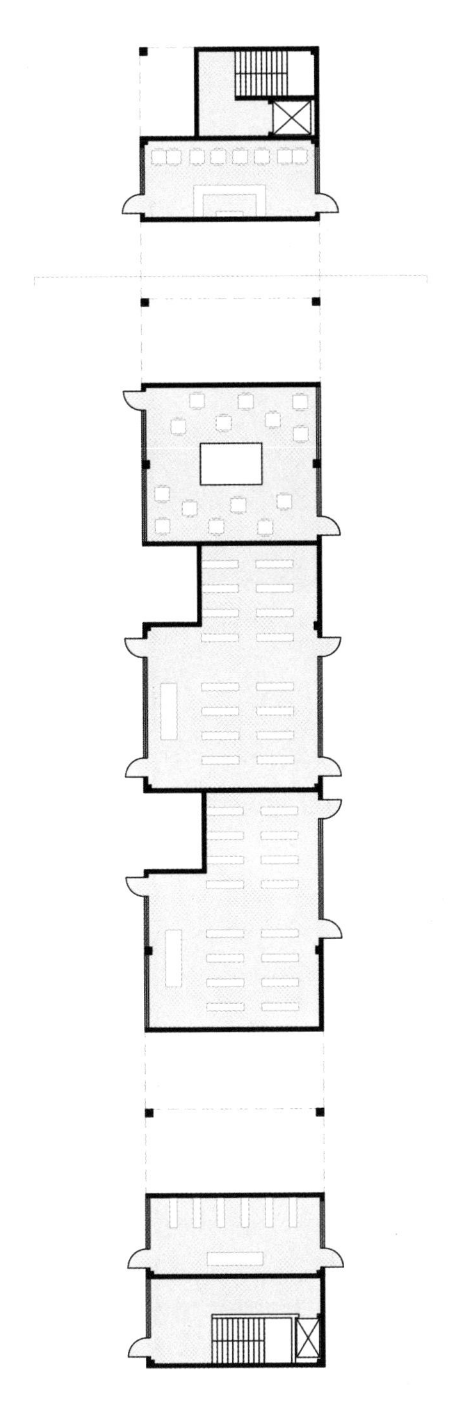

商业建筑首层平面
Commercial Building - First Floor

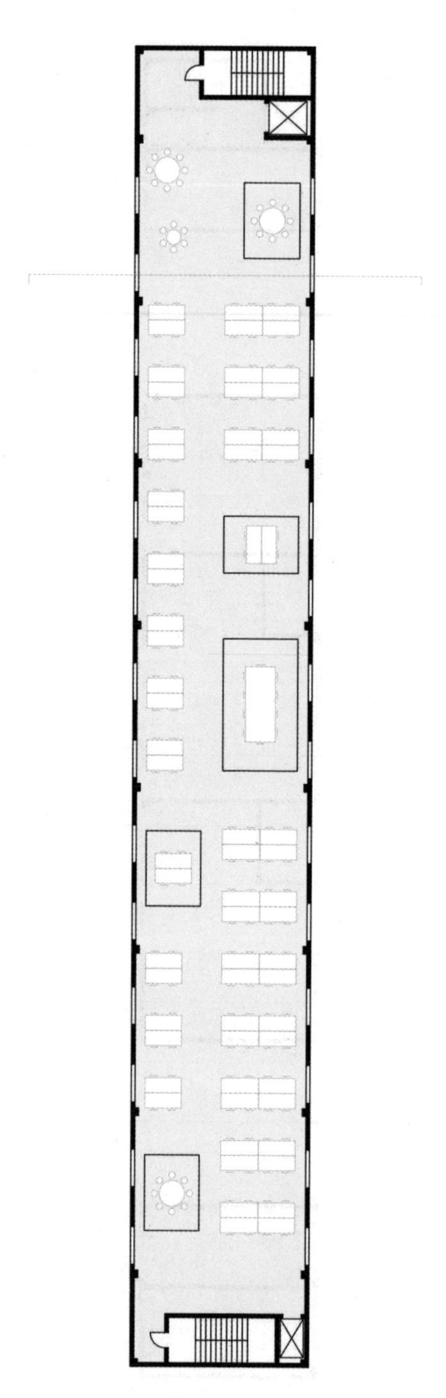

商业建筑共享办公空间
Commercial Building - Co-Working Space

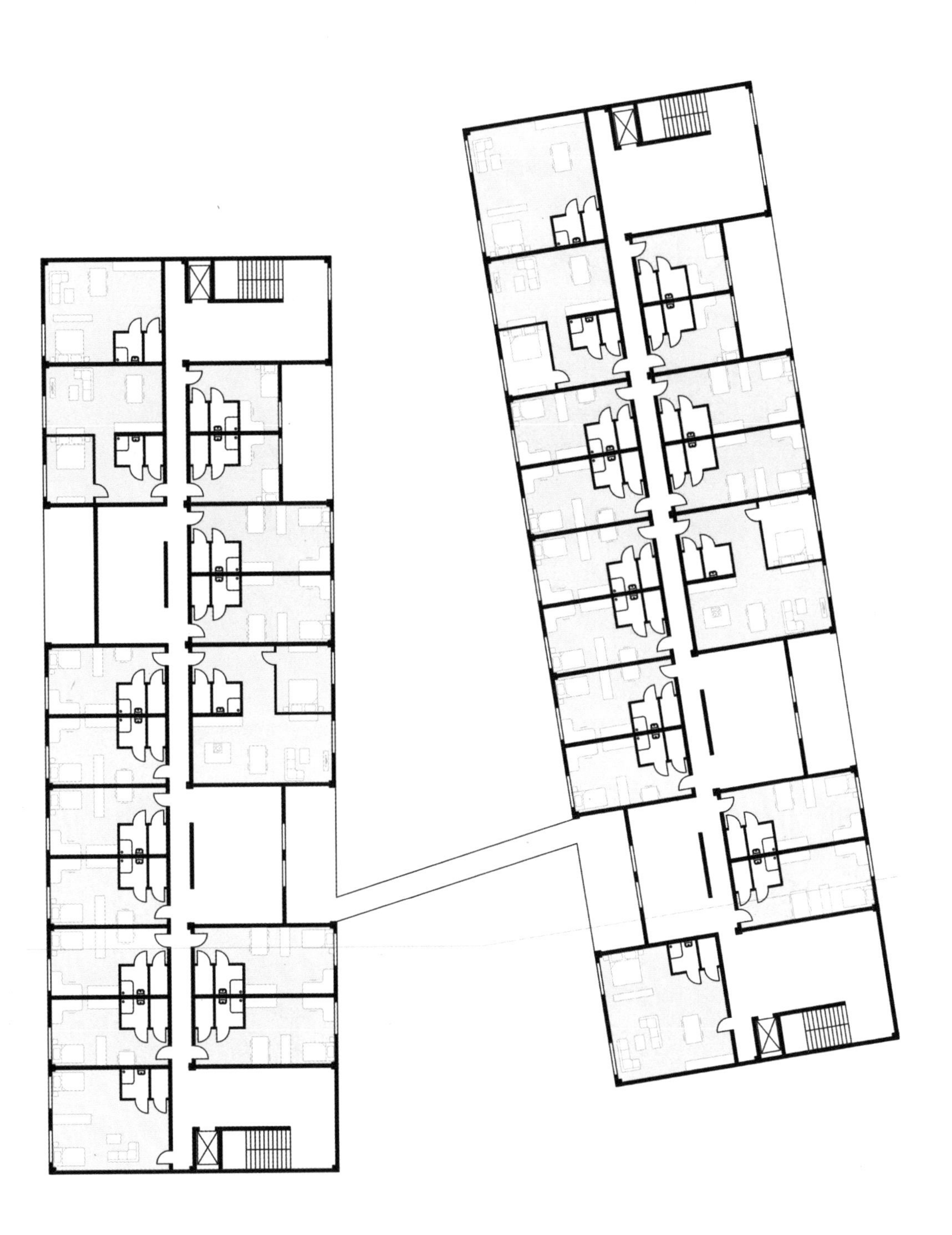

宿舍标准层平面
Dormitory - Regular Floor Plan

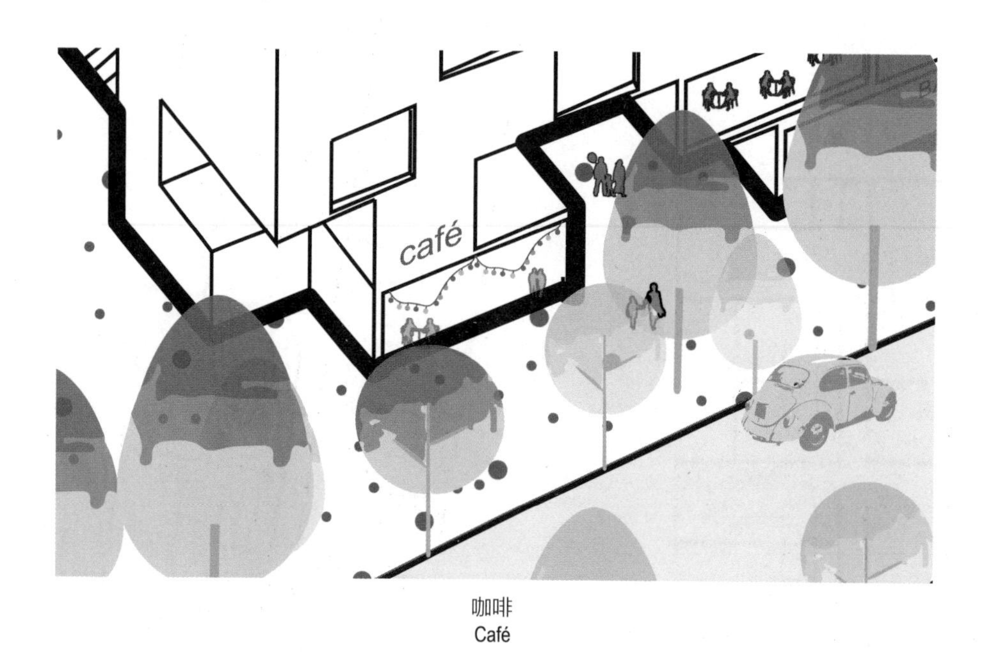

咖啡
Café

日用品商店
Daily Use Shops

屋顶活动空间
Roof Top Activities

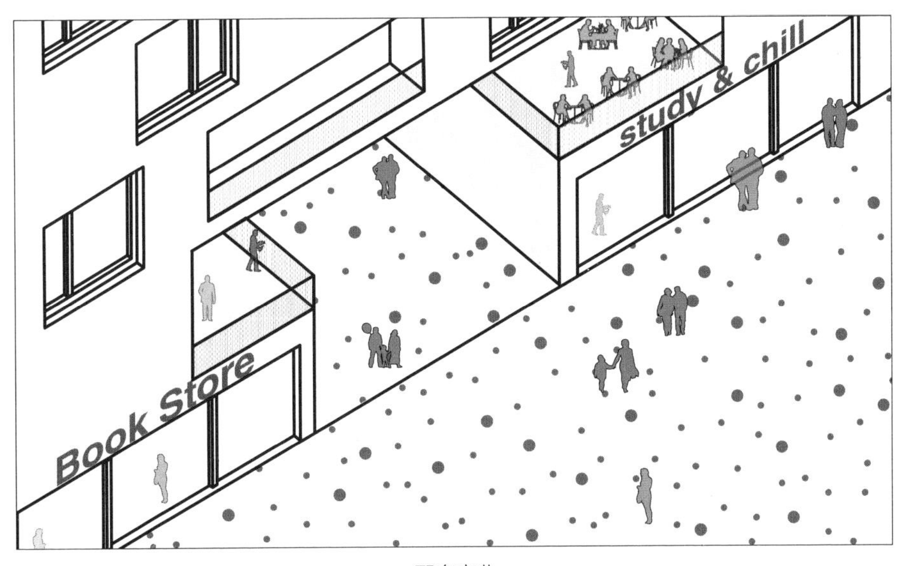

配套商业
Stores according to users

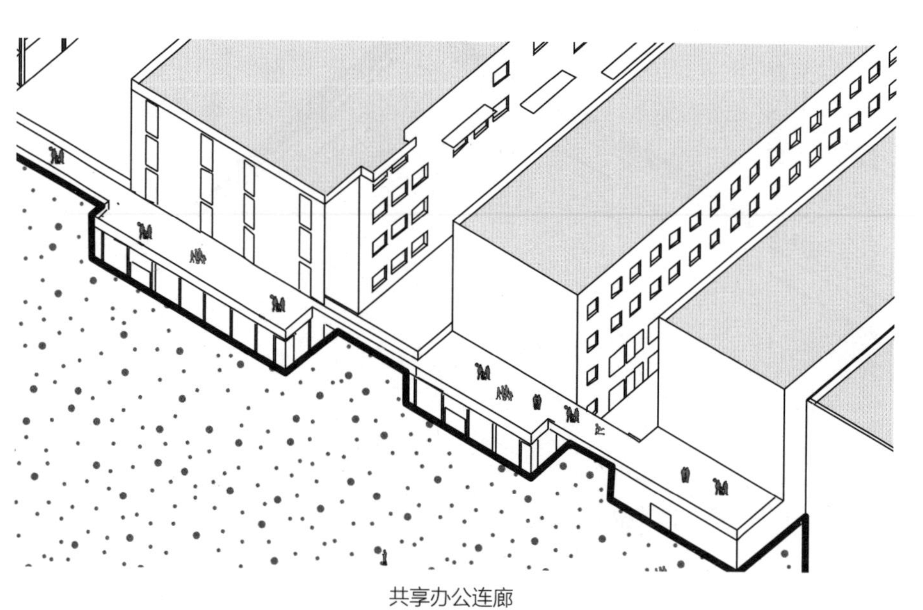

共享办公连廊
Co-Working Connection

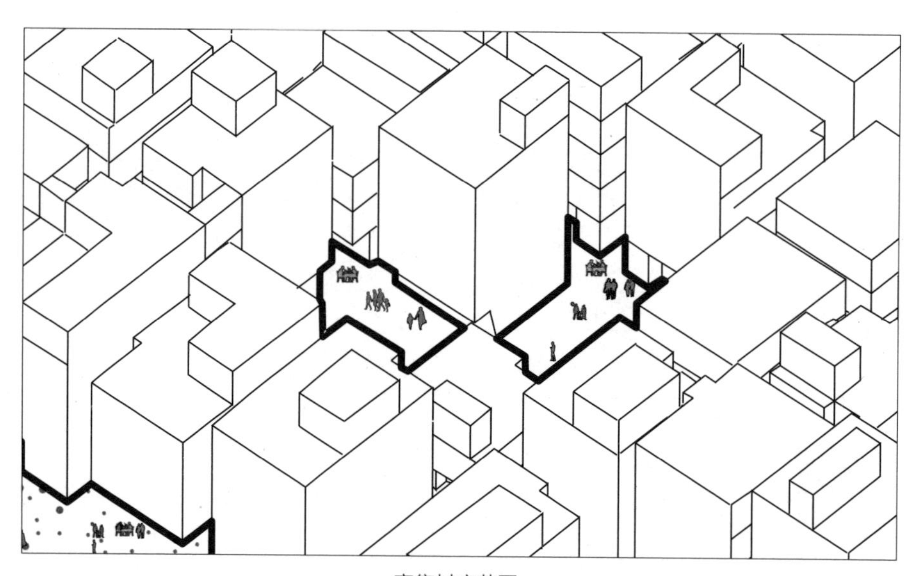

密集村庄花园
Dense Village Gardens

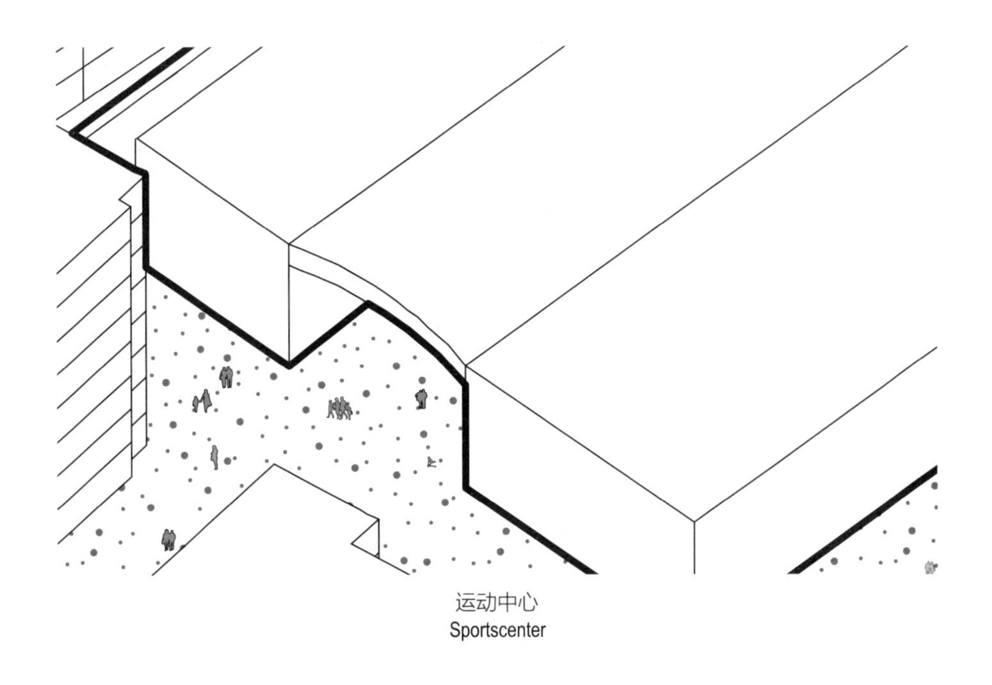

运动中心
Sportscenter

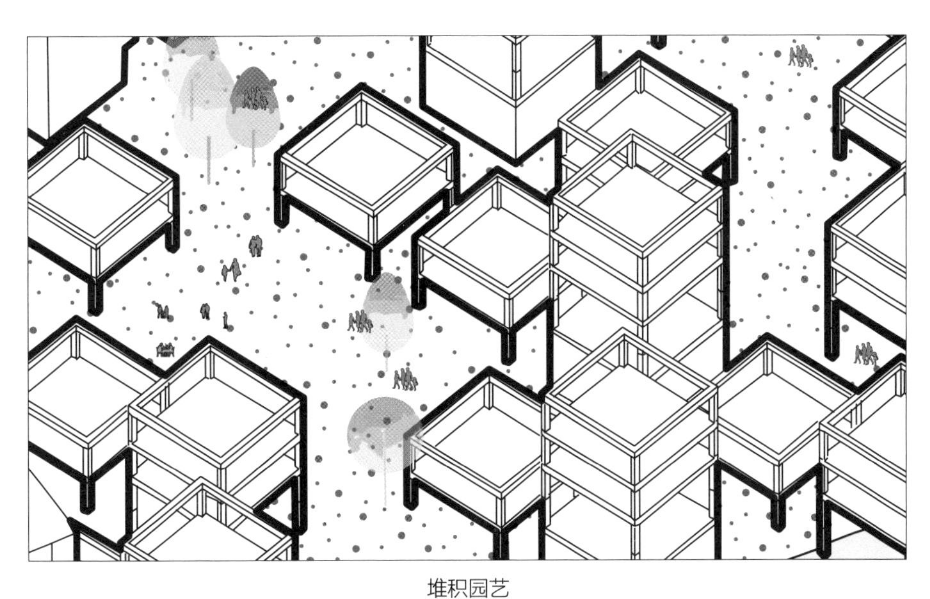

堆积园艺
Stacked Gardening

6.2 滨河空间

RIVER FRONT

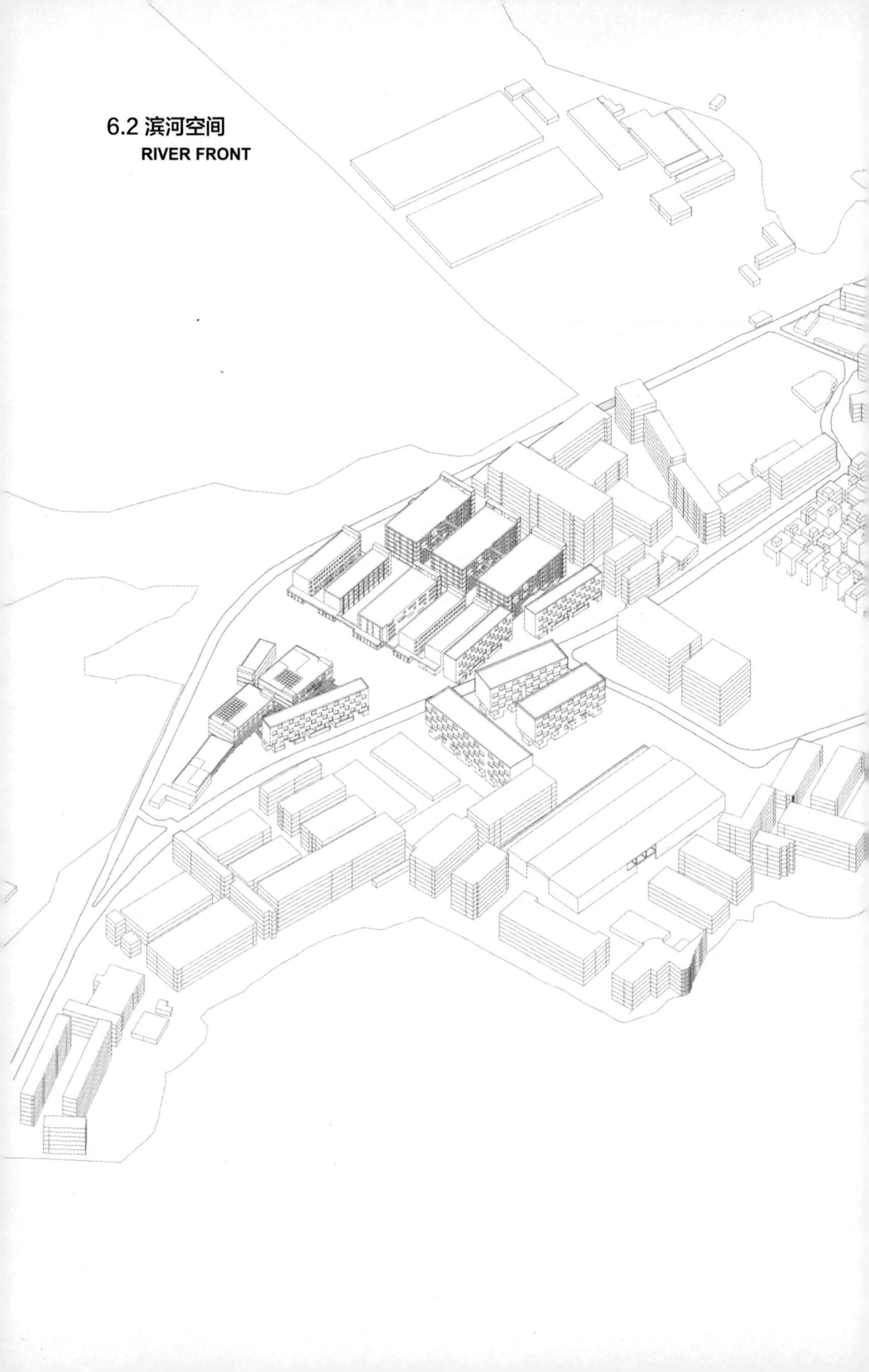

滨河空间轴测图
Riverfront axonometric

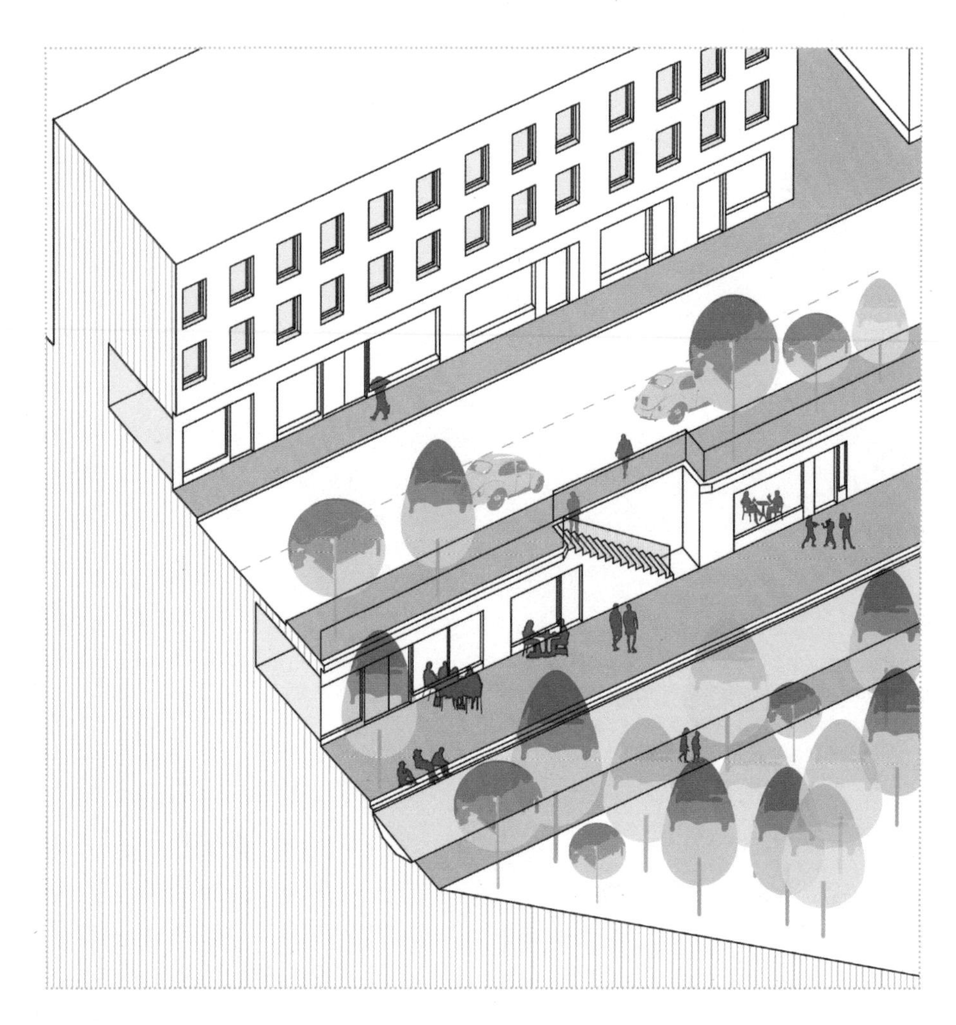

河岸断面
Riverfront section

滨河开发

目前的河滨区主要是一片绿地，在周边河流与人的住所之间形成绿色走廊。在我们的概念中，这种情况具有很大的潜力，将居民更多地与自然联系起来，并创建一个小规模的地方，在那里他们可以互相交流。在我们对芝加哥和Cheonggecheon河滨规划进行研究之后，我们提出了一个具有两层城市空间的概念，以拉近河流与人距离。

Riverfront development

The riverfront in current situation is mainly a green space, which creates a green corridor between the river and the residence. In our concept this situation has a lot of potentials, to connect the inhabitatnts more to the nature, and create a small scale place, where people can gather for communication.
After our research on the Chicago and Cheonggecheon riverfront plan, we propose a concept with two layers of urban space, to bring the river closer for people.

河岸现状
Current situation of the riverfront

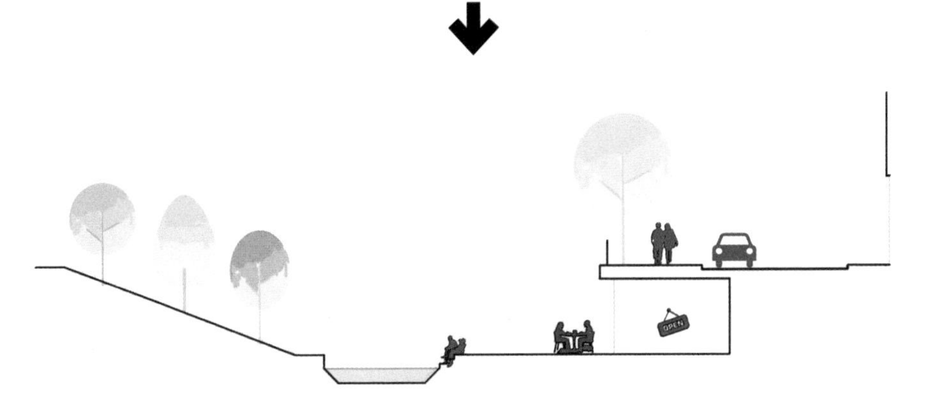

河滨餐厅与自然相连
Riverfront with restaurants and visual connection to nature

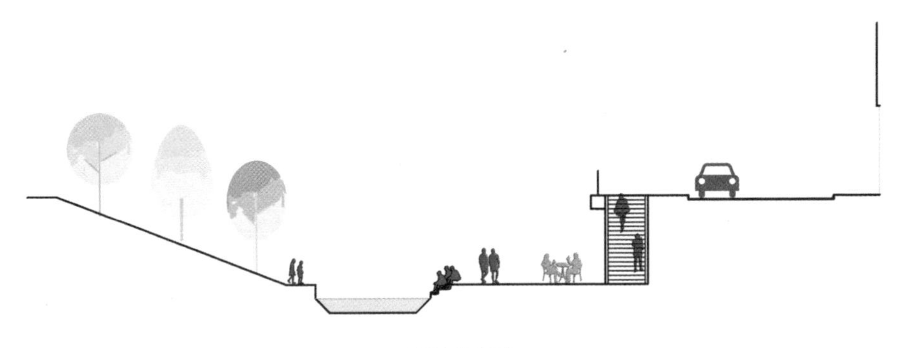

河滨步行道
Riverfront, access to the walkway

改造策略
RENEWAL STRATEGIES

滨河空间透视图
Riverfront perspective

6.3 街道界面
STREET INTERFACE

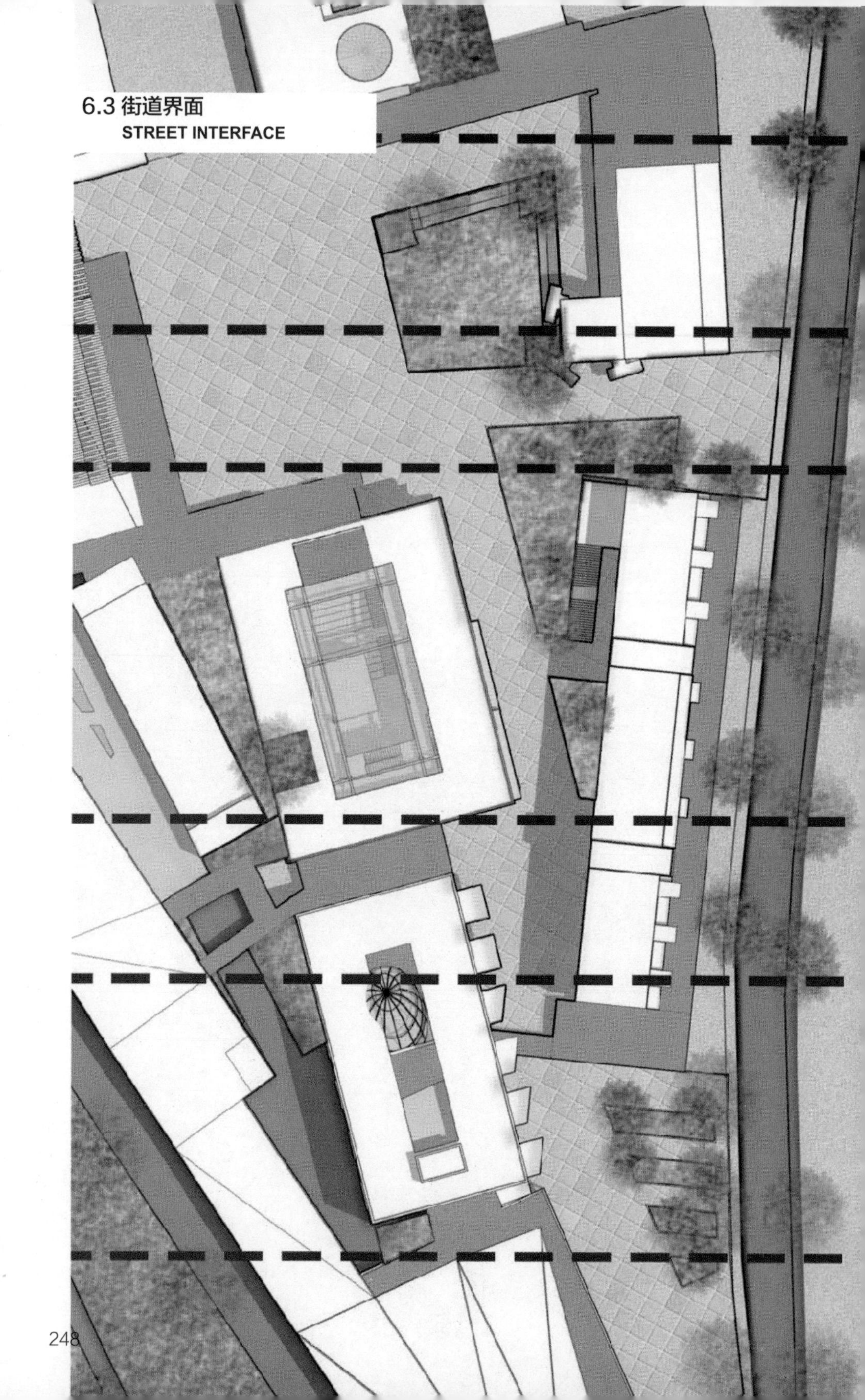

街道剖面
Street sections

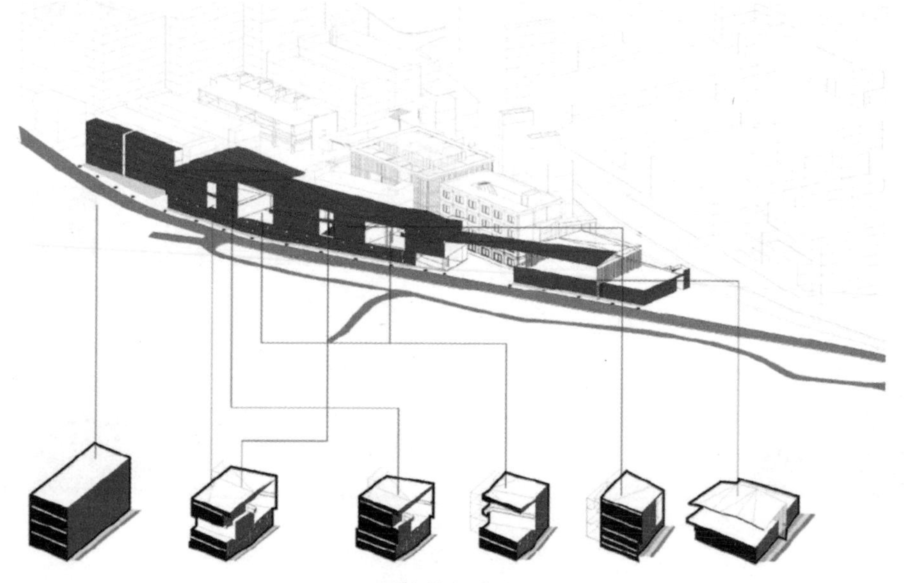

过境道路界面
Transit road interface

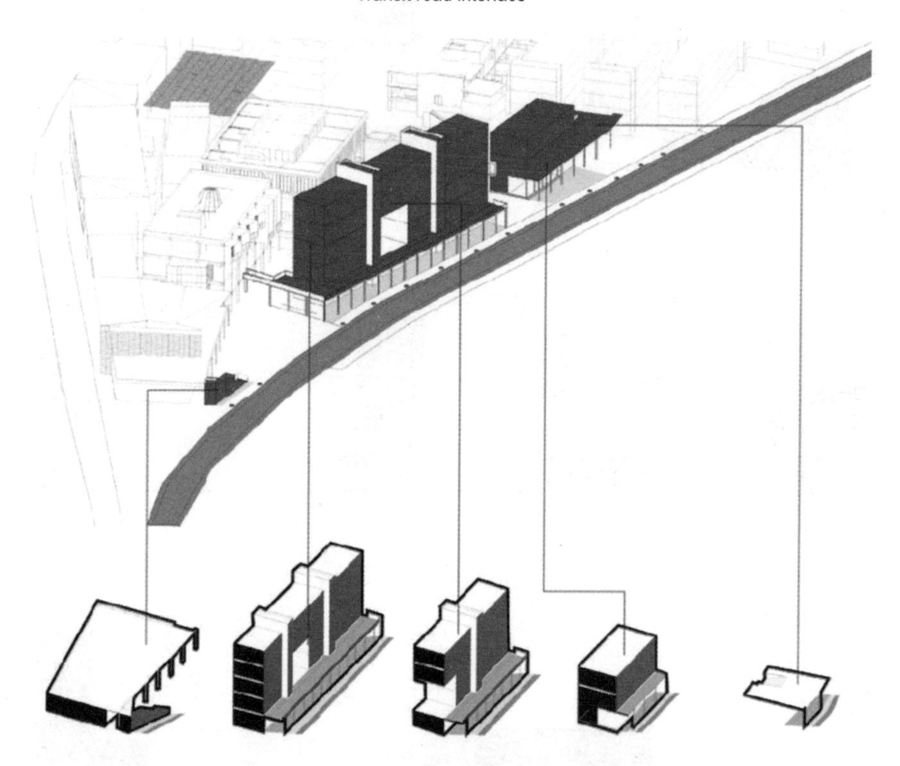

沿村主街界面
Along the village main street interface

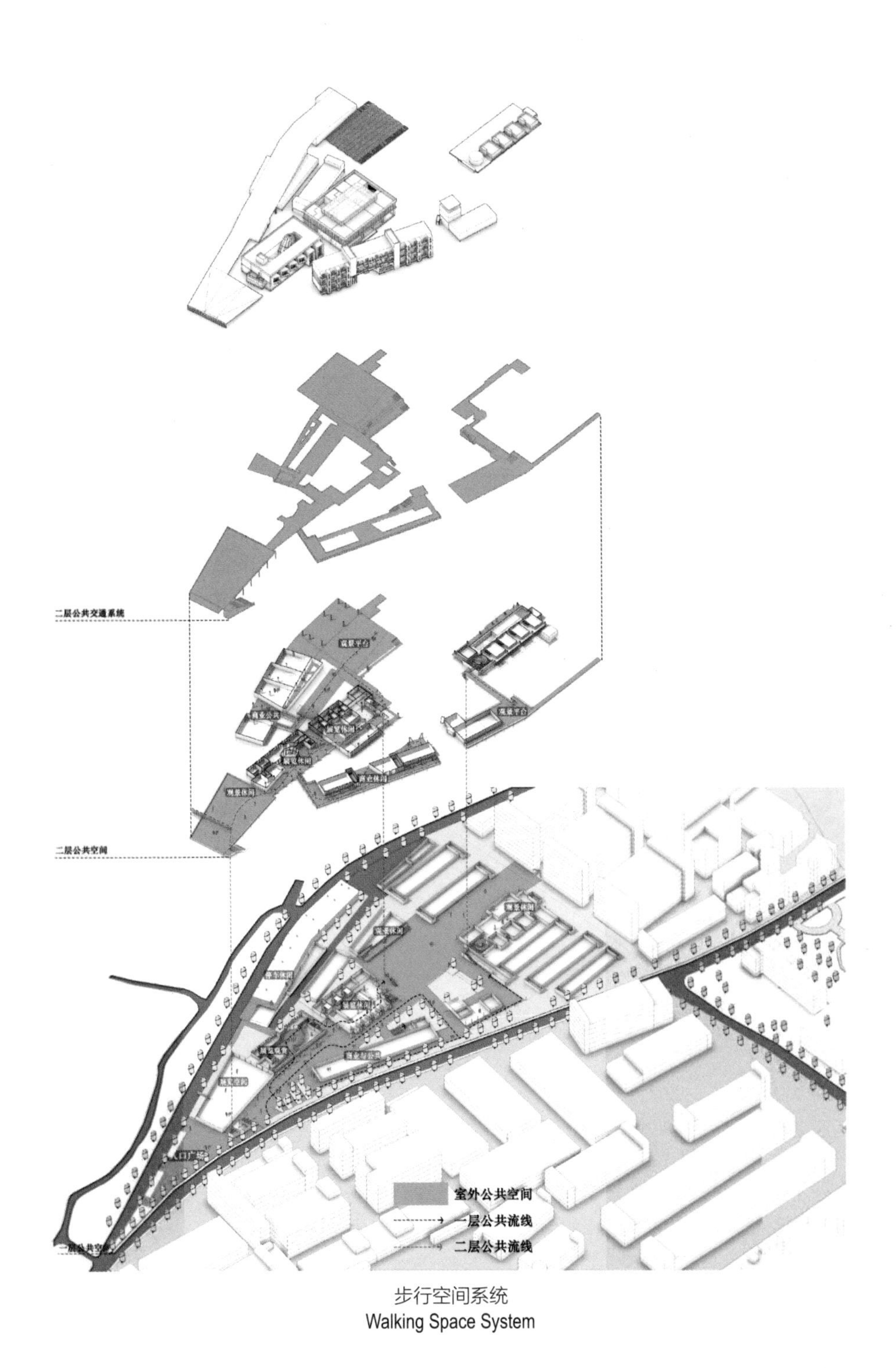

步行空间系统
Walking Space System

6.4 堆叠花园
STACKED GARDEN

堆叠花园轴测图

Stacked garden axonometric

堆叠花园透视图

Stacked gardens perspective

堆叠花园透视图 II

Stacked gardens perspective drawing II

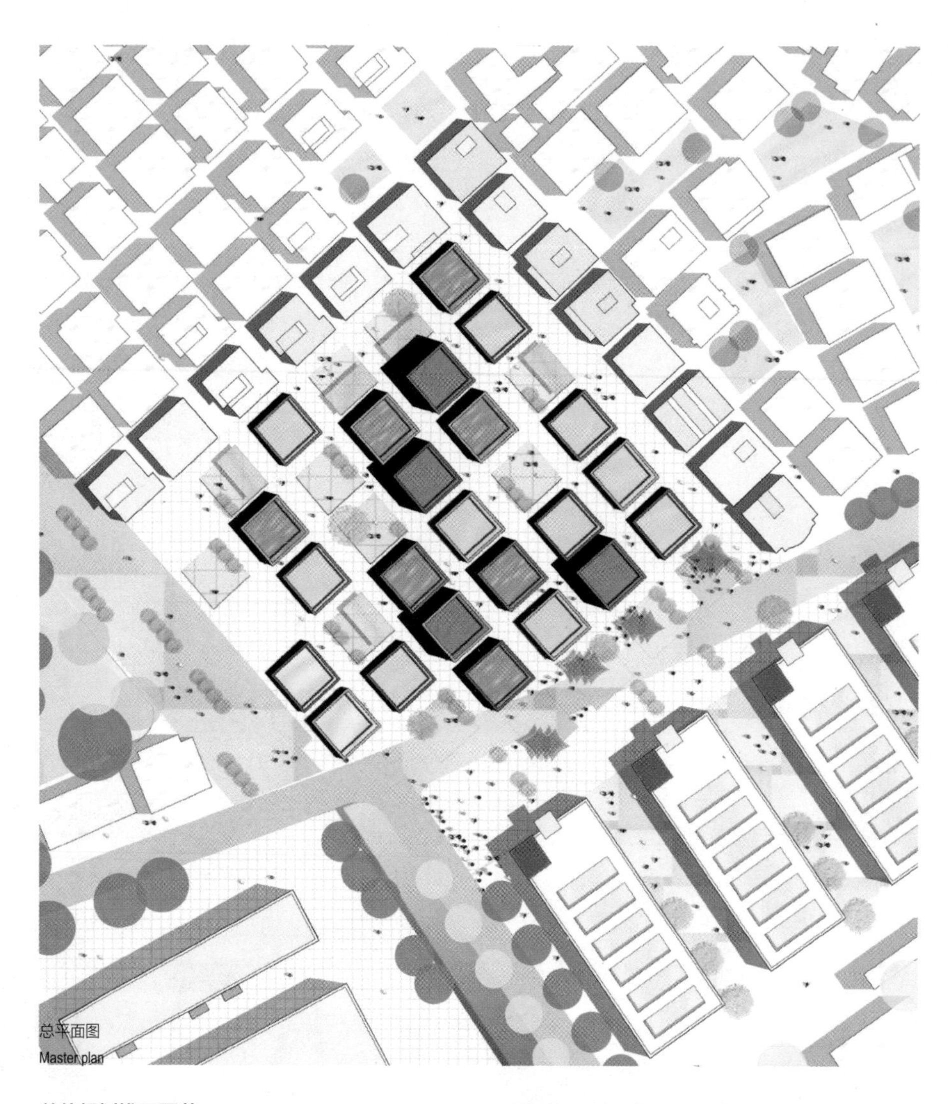

总平面图
Master plan

总体规划堆积园艺

广场中间的堆叠花园旨在为地面层的所有观众开放，并将白领和蓝领混合在一起。附近的工人可以负责花园和地面设施，并为入住已改造工厂的年轻技术人员提供产品和服务。技术人员必须进入工人社区以满足他们的日常需求，但也可以在此进行休闲活动。另一方面，蓝领工人们必须前往原工厂内的屋顶花园维护花园环境，同时他们也有机会在年轻企业中获得工作机会。

Masterplan Stacked Gardening

The stacked gardens in the middle of the square aim to be open for all audiences in the ground floors and to mix both the young academics and the proletarians. The workers from the neighbourhood could be in charge of the gardens as well as the groundfloor facilities and offer their products and services to the young professionals from the readapted factories. The professionals would necessarily have to go to the workers neighbourhood to cover their daily needs but could also use it for leisure activities. On the other hand the workers have to go to the rooftop gardens of the former factories to keep them alive and possibly also to take advantage of some working possibilities in the young enterprepreneurs' new businesses.

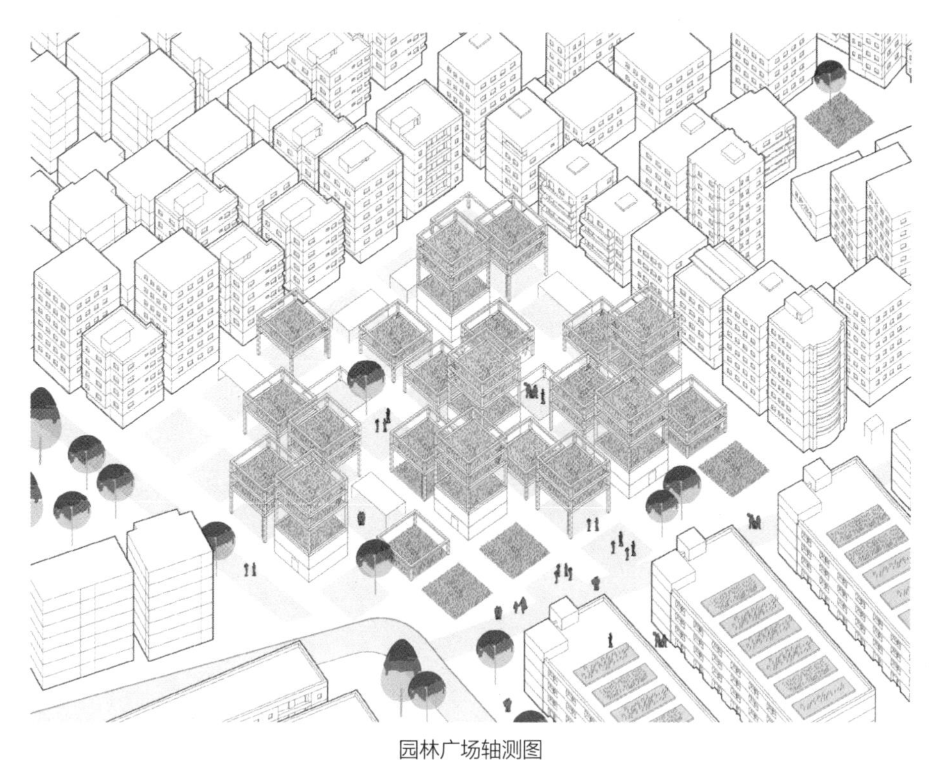

园林广场轴测图

Axonometric drawing of the square with stacked gardens

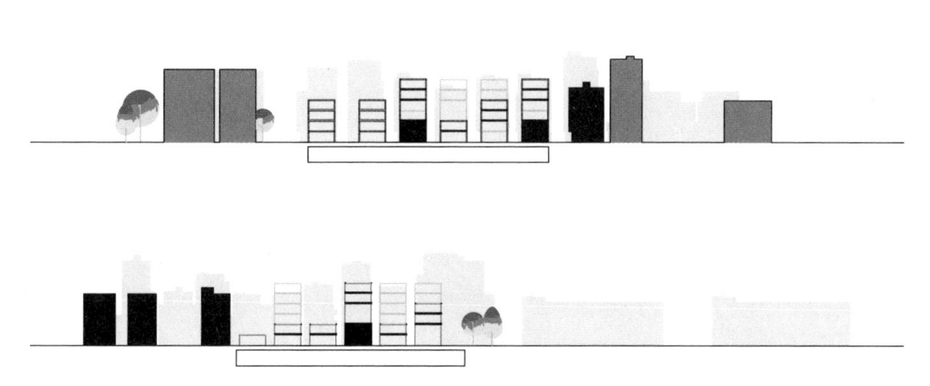

广场上园艺堆叠

Section of the square with stacked gardens

建筑物的轴测
Axonometric drawing of the readapted buildings

工厂重新适应了创意中心

旧厂房要重新利用。它们的结构可以保留，而室内空间应重新设计和改造。首层可以设置较大空间尺度的工作坊。楼上各层，除了可以改造为尺度更小的工作坊、工作室和办公空间，还可以作为公寓等。每层楼都有一个公共厨房，而顶层应该作为公共广场、屋顶共享花园。项目的目标群体可以是年轻的毕业生、艺术家和刚刚进入社会的其他专业人士等。屋顶共享花园也可以提供给在社区另一侧的务工人员使用。楼下的尺度多样的小广场则更适合居住于此的周边居民使用。因此，改造后的首层空间更像似一个公共的社区空间和商业空间，而广场则变成更接近花园的感觉。

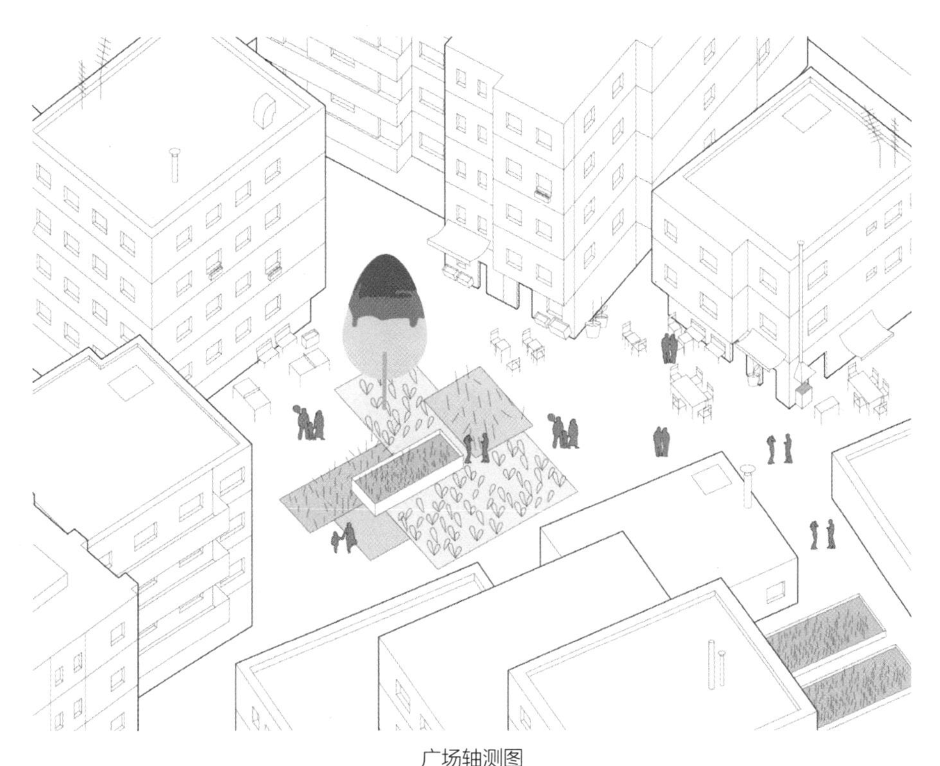

广场轴测图
Detail axonometric drawing of one of the small squares

Factories readapted to creative hubs

The former factories are to be reused. Their structural elements shall be preserved, the interior space redesigned and readapted. The ground floor should host workshops on a larger scales. The upper floors should host smaller scale workshops, studios and offices on the one side, and small appartments and rooms on the other side. Every floor should have a communal kitchen. The rooftop is to be used as the public squares - as gardening space. The target group are young graduates, artists and other professionals who seek a practical but inexpensive working and living space. The roof gardens are to be used by the workers in the area on the other side of the street. The small squares should be more or less a downscaled version of the big square, just that they differ in size and that they don't aim the whole village (or wider) as a target group, but rather the direct neighborhood. The groundfloor are likewise having communal or commercial use, the square hosts a garden.

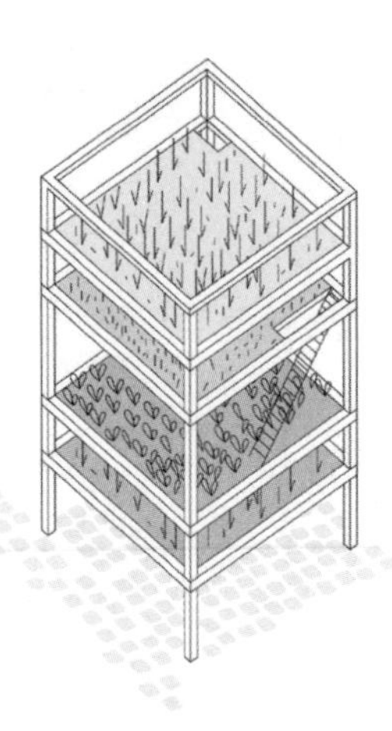

底层开放空间
Groundfloor as open space

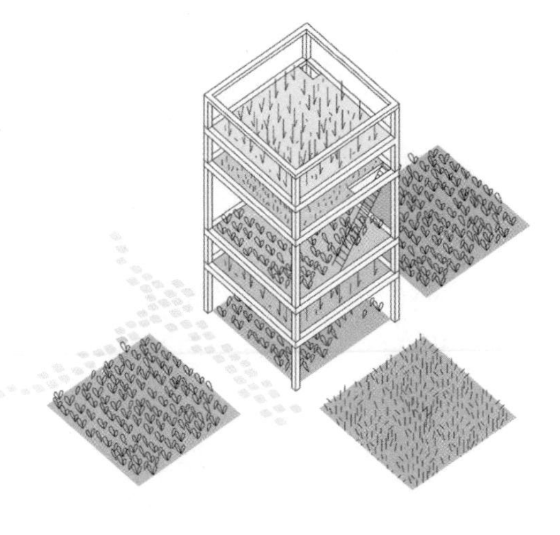

底层花园
Groundfloor as garden

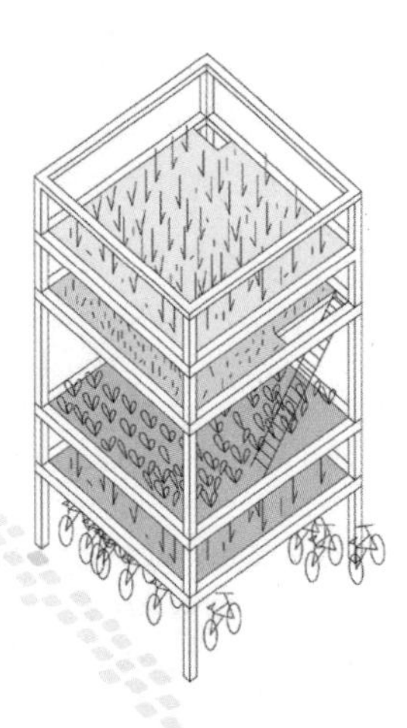

底层自行车停车场
Groundfloor as bike storage

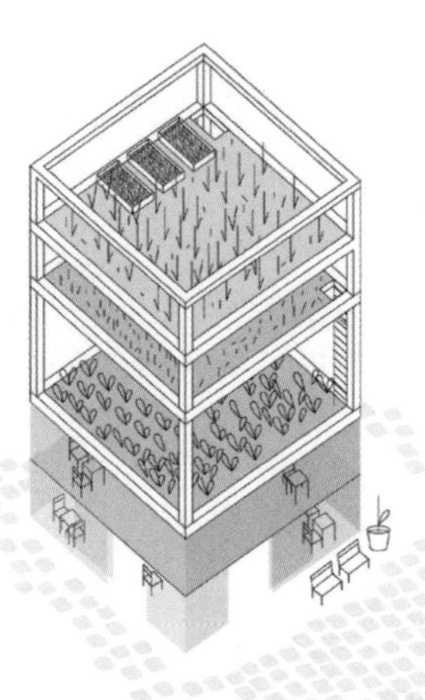

底层商业
Groundfloor as commercial use

塔楼的地面使用情况

每个堆叠式花园塔楼的地面使用都有多种可能。上层一般用作园艺空间。 地面层可以用于与园艺相关的活动，但它们应该主要用作可公开进入的开放空间，自行车存放处，咖啡店等。地面设施应该服务于附近人群。

Groundfloor usage of the towers

The groundfloor usage of each stacked garden tower has the possibility to defer.The upper levels are always used as gardening space. While the groundfloors can be used for activities related gardening as well, they should mainly act as publicly accessible open spaces, bike storages, coffeshops etc. The facilities on the groundfloor should serve the people of the neighbourhood.

6.5 水资源净化

WATER PURIFICATION

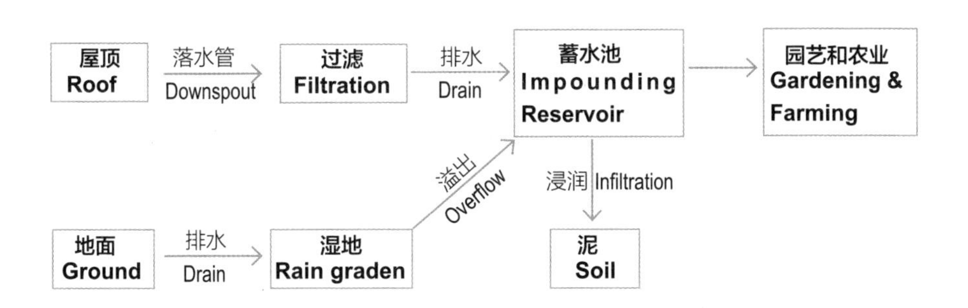

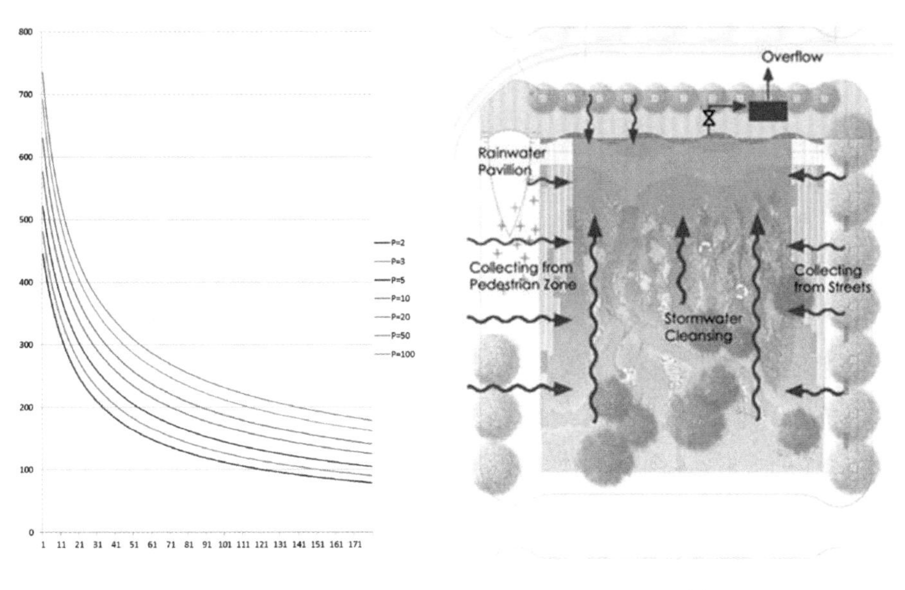

降雨量

Rainfall

雨水循环

Rainwater circulation

水资源净化流程

The process of water purification

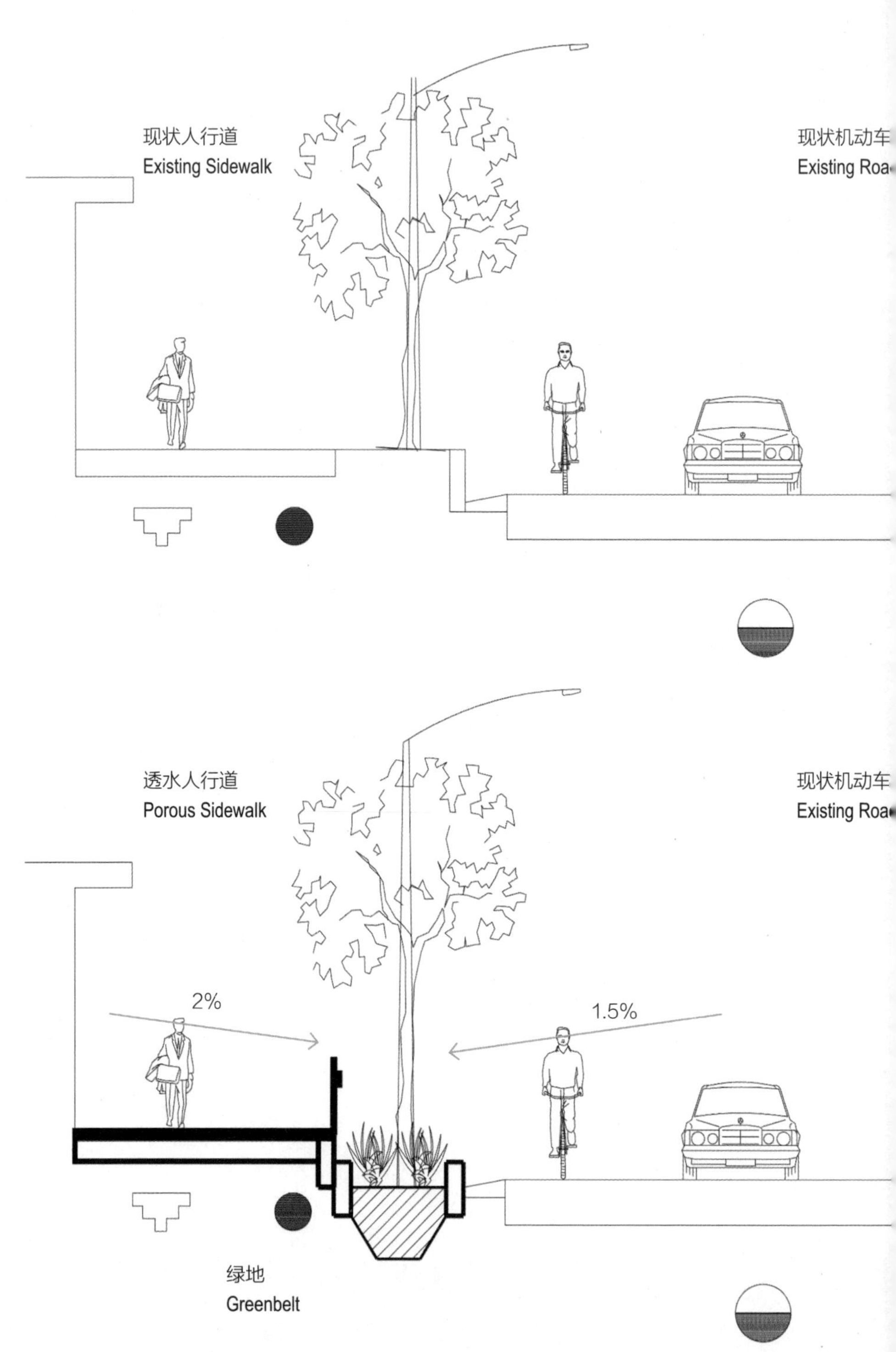
现状人行道
Existing Sidewalk
现状机动车
Existing Roa
透水人行道
Porous Sidewalk
现状机动车
Existing Roa
2%
1.5%
绿地
Greenbelt

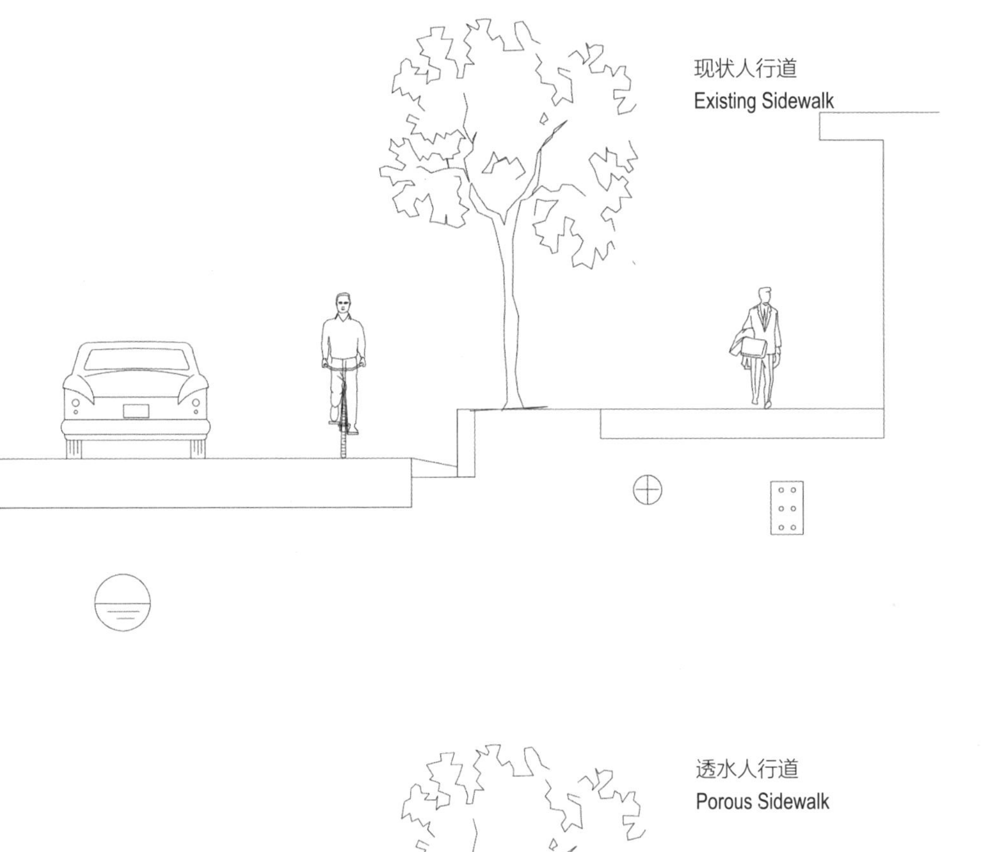
现状人行道
Existing Sidewalk

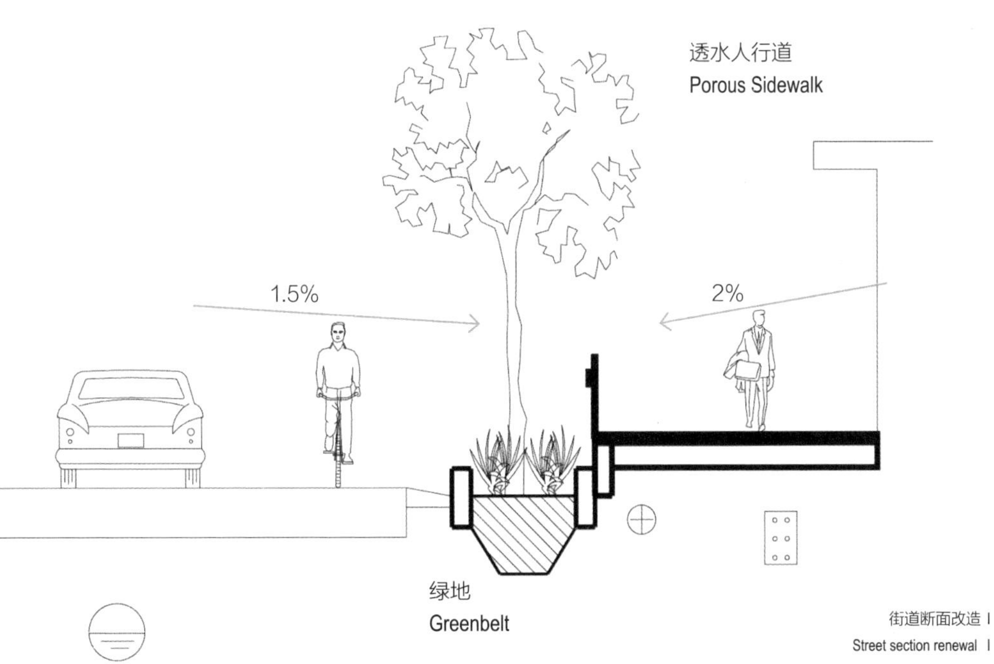
透水人行道
Porous Sidewalk
1.5%
2%
绿地
Greenbelt

现状绿地
Existing Hill

现状河道
Existing River

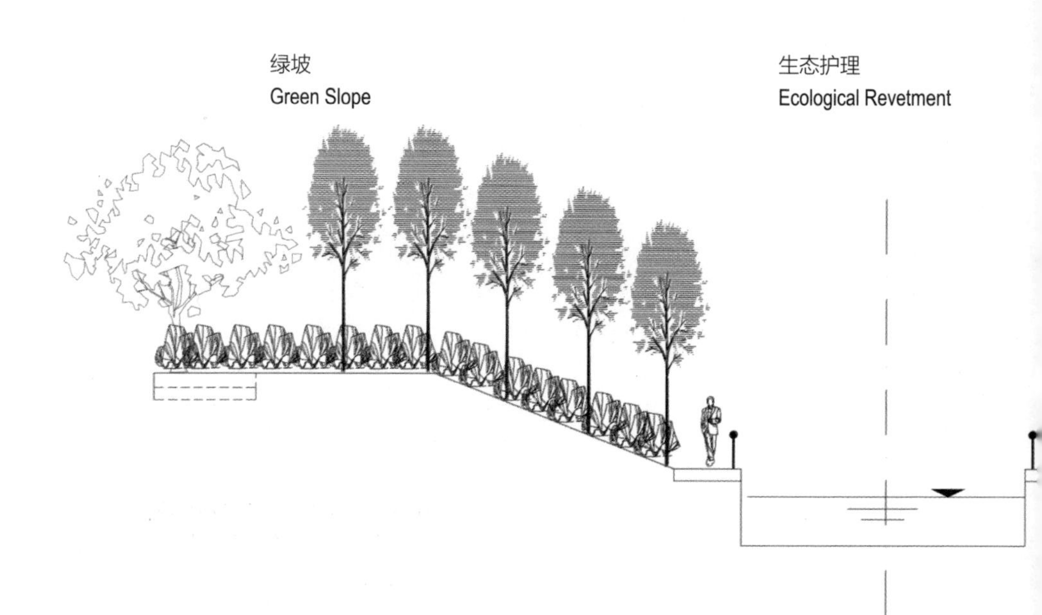

现状车道与人行道

Existing Roadway and Sidewalk

透水人行道

Porous Sidewalk

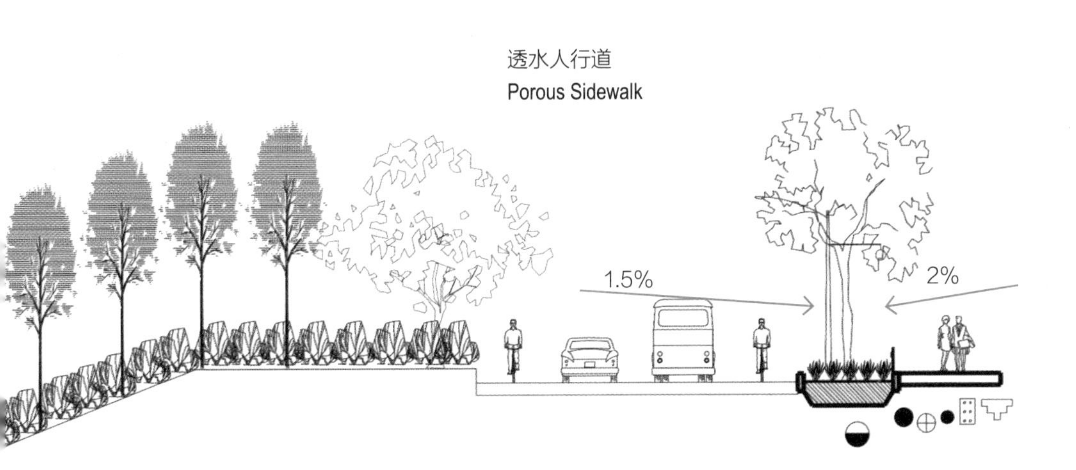

住宅立面绿化
Residential facade greening

生态种植
Ecological planting

生态种植
Ecological planting

生态种植
Ecological planting

图书在版编目（CIP）数据

城中村的新生 = Rebirth of The Urban Village : 英汉对照 / 杨镇源等著 . — 北京 : 中国城市出版社 , 2020.12

（城市设计研究）

ISBN 978-7-5074-3317-3

Ⅰ . ①城… Ⅱ . ①杨… Ⅲ . ①居住区 — 旧城改造 — 深圳 — 英、汉 Ⅳ . ① TU984.265.3

中国版本图书馆 CIP 数据核字 (2020) 第 242744 号

责任编辑：滕云飞
美术编辑：朱怡勰
责任校对：焦　乐

城市设计研究

城中村的新生

REBIRTH OF THE URBAN VILLAGE

杨镇源 肖　靖（深圳大学） 李凌月 许　凯 孙彤宇（同济大学） 唐　斌（东南大学）
［奥］莫拉登·亚德里奇（Mladen Jadric）（维也纳技术大学） 著

*

中国城市出版社出版、发行（北京海淀三里河路 9 号）
各地新华书店、建筑书店经销
临西县阅读时光印刷有限公司印刷

*

开本：787 毫米 ×1092 毫米　1/16　印张：17¼　字数：378 千字
2021 年 9 月第一版　　2021 年 9 月第一次印刷
定价：**226.00** 元

ISBN 978-7-5074-3317-3
（904291）